WILD DISCOVERY

GUIDE TO YOUR

DOG

◆　◆　◆

GUIDE TO YOUR

DOG

UNDERSTANDING AND CARING
FOR THE WOLF WITHIN

◆ ◆ ◆

FOREWORD BY
ELIZABETH MARSHALL THOMAS

DISCOVERY BOOKS
New York

DISCOVERY COMMUNICATIONS

Founder, Chairman, and Chief Executive Officer:
John S. Hendricks
President and Chief Operating Officer: Judith A. McHale
President, Discovery Enterprises Worldwide: Michela English
Senior Vice President, Discovery Enterprises Worldwide:
Raymond Cooper

DISCOVERY PUBLISHING

Vice President, Publishing: Natalie Chapman
Editorial Director: Rita Thievon Mullin
Senior Editor: Mary Kalamaras
Editor: Maria Mihalik Higgins

DISCOVERY CHANNEL RETAIL

Product Development: Tracy Fortini
Naturalist: Steve Manning

Discovery Communications, Inc., produces high-quality television programming, interactive media, books, films, and consumer products. Discovery Networks, a division of Discovery Communications, Inc., operates and manages the Discovery Channel, TLC, Animal Planet, Travel Channel, and Discovery Health Channel.

Wild Discovery Guide to Your Dog was created and produced by St. Remy Media Inc. for Discovery Publishing.

Library of Congress Cataloging-in-Publication Data
Wild discovery guide to your dog : understanding and caring for the
    wolf within / foreword by Elizabeth Marshall Thomas
          p.   cm.
    Includes bibliographical references
    ISBN 1-56331-805-9
    1. Dogs. 2. Dogs—Behavior. I. Discovery Communications, Inc.
SF426.W56   1999
636.7—dc21   99-31016
                                                  CIP

Random House website address: www.atrandom.com
Discovery Communications website address: www.discovery.com

Printed in the United States of America on acid-free paper.
9 8 7 6 5 4 3 2 1
First Edition

CONSULTANTS

Elizabeth Albanis, *certified in canine obedience instruction, is co-owner of the Montreal, Canada, consultation service, Canine Education Plus, which specializes in dog training and training problems. She has worked as a private dog trainer for the past six years, and at the Montreal SPCA for the past four.*

Linda Goodloe, Ph.D., *is a certified applied animal behaviorist. Based in New York City, she has a private clinical and consulting practice for behavior problems of companion animals.*

Jack Grisham *is the director of animal management at the Oklahoma City Zoological Park. He is also involved with the American Zoo and Aquarium Association, Species Survival Plan Management groups, and the International Union for the Conservation of Nature (I.U.C.N.) Canid Specialist Group.*

Debbie Groves, *co-owner of the Montreal, Canada-based Canine Education Plus, a consultation service for dog training and training problems, has been an adoption consultant and dog temperament evaluator at the Montreal SPCA for the past twelve years.*

Nigel Gumley, D.V.M., *is a partner in the Alta Vista Animal Hospital in Ottawa, Canada. He is a past-president of the Ontario Veterinary Medical Association, and is currently on the National Issues Committee of the Canadian Veterinary Medical Association.*

Connie Jankowski *is a California-based pet care expert whose work has included the development of breed descriptions that appeared in Dogs USA magazine. Over the years, she has worked with both veterinarians and pet-related businesses on various projects regarding pets.*

Margaret E. Lewis, Ph.D., *is an assistant professor of biology at the Richard Stockton College of New Jersey. She specializes in the behavior, ecology, and evolution of carnivores.*

Ravi Seshadri, D.V.M., D.A.B.V.P., *is in private veterinary referral practice in Southern California. He has completed a residency in emergency and critical care, and frequently lectures and teaches courses to vets and their support staff.*

CONTRIBUTOR

Elizabeth Marshall Thomas, *an anthropologist, is the author of the best-selling books* The Hidden Life of Dogs *and* The Tribe of Tiger. *She lives in New Hampshire with her three dogs, nine cats, and four parrots. Her long list of publications includes a novel,* Reindeer Moon.

ST. REMY MEDIA

President: Pierre Léveillé
Vice-President, Finance: Natalie Watanabe
Managing Editor: Carolyn Jackson
Managing Art Director: Diane Denoncourt
Production Manager: Michelle Turbide
Director, Business Development: Christopher Jackson

Senior Editor: Heather Mills
Art Director: Odette Sévigny
Assistant Editors: Stacey Berman, Jim Hynes
Researcher Writers: Ida W. Estep, Pierre Home-Douglas
Photo Researcher: Linda Castle
Indexer: Linda Cardella Cournoyer
Senior Editor, Production: Brian Parsons
Systems Director: Edward Renaud
Technical Support: Roberto Schulz, Jean Sirois
Scanner Operators: Martin Francoeur, Sara Grynspan

· CONTENTS ·

Top: Shetland sheepdogs
Left: Coyotes

·FOREWORD·

"How much better we would be as dog owners if we knew who dogs really are and understood what motivates them."

◆

ELIZABETH MARSHALL THOMAS

Why do dogs enjoy such popularity among members of our species? Perhaps because they possess many of the qualities to which we also aspire, attributes such as intelligence, strength, loyalty, and a sense of purpose, to name but a few. These characteristics kept the family groups of their wolf ancestors successful and cohesive. In short, over the eons these traits have helped the wolves survive. And as the owners of dogs, we now use these splendid qualities to our advantage, even as we take them for granted. We feel, for instance, that the loyalty of a dog is our due, and we don't realize that the dog's wish to be loyal springs from a young wolf's need to keep together with his parents even after he's grown. He remains with them to help them hunt, to fight by their sides if necessary, and to help them raise his infant siblings, sometimes feeding the pups with food from his own stomach, food which he may need himself. Dogs would do all these things for their owners, if they knew how.

Although many of us appreciate the loyalty of dogs, we may also exploit it, perhaps leaving a young dog alone for a long time in a crate. His wish to be with his group is strong—wolves must keep their groups together—and the young dog cries because he is unable to do what his ancestry tells him he must do: stay with the others. His crying annoys us, and we discipline him.

How much better we would be as dog owners if we knew who dogs really are and understood what motivates them. To do so, however, we must benefit from many different kinds of information, from their history as a species to their physiology, from training methods to their sociology as pets. To compile a comprehensive book about an animal as complex as a dog is no easy task, but fortunately for dogs, and also for us all, the editors of this work have done so to perfection. Here is no reiteration of the old cliches—here is new material combined with old wisdom, as informative as it is entertaining.

Elizabeth Marshall Thomas
Author of *The Hidden Life of Dogs*

Opening photographs:
Page 2: Gray wolves
Pages 6-7: Gray wolves
Pages 8-9: Golden retriever
Pages 10-11: Red foxes
Pages 12-13: Beagle
Opposite: Gray wolf

DOG
PRIMER

◆   ◆   ◆

**"Who keeps company with the wolf will
learn to howl."**

◆

16TH-CENTURY PROVERB

TAMING THE
· WOLF ·

Like this band of modern-day gray wolves, prehistoric wolf packs must have inspired both fear and awe in those who were involved in their domestication. Although no domestic dog breed's ancestry can be traced to a single sub-species of prehistoric wolf, wolves from North America, Europe, middle Asia, and the Far East provided enough variation in body and coat types to create the gene pool that gave rise to today's vast array of dog types.

Opposite: Though this gray wolf may physically resemble only a few of today's domestic dogs, most notably the Siberian husky and Alaskan Malamute, there is still a little of the wolf's nature in each of the more than 400 breeds. Despite their shyness, wolves were amenable to domestication because of their innate sociability and their understanding of dominant-submissive relationships, found in both wolf packs and human families.

Overleaf: Timber wolves

To most people, the fearsome wolf of some fact and much fable seems too far removed from today's domestic dog to suggest anything more than a distant kinship. After all, the vast majority of dogs are gentle and usually crave human companionship. Wolves, on the other hand, are notoriously shy and can be extraordinarily vicious when cornered or confronted. And yet the fact remains: The progenitor of man's best and oldest friend is none other than the legendary creature who still symbolizes all that is wild and unharnessed in the world.

How were humans able to turn the wolf's unbridled free spirit into the lovable and loyal domestic dog of today? Undoubtedly hunger played a major role in breaking the ice. The same scavenging behavior that we find so inexplicable in today's well-fed pets most likely prompted some of the braver wolves to creep close to our forefathers' campfires in search of scraps, setting the stage for an enduring alliance. But there were other factors at work that made domestication possible, some as deep-seated as the psychological makeup of both man and beast.

MAN MEETS WOLF

Because wolf habitat ranged from Southern Asia to the arctic tundra, wolf and man had a chance to coexist long before domestication began. But whether or not human beings actively sought to tame the wolf is a matter of much debate. In fact, some argue that wolves actually domesticated themselves by choosing to live near man. They were certainly clever enough to connect human encampments with food. But even before full-fledged domestication began, a loose association—or at the least a spirit of mutual tolerance—must surely have existed as wolves and humans lived and hunted in close proximity. No doubt, behavioral similarities also helped foster the relationship. Both wolves and humans are highly social beings; both live in hierarchically structured packs or families that hunt co-operatively and share in the raising of their young.

Nobody knows for sure when humans and wolves first began to interact. Some scientists suggest that it may have occurred from 40,000 to 100,000 years ago. It would take at least that long, they say, for the genetic differences that exist between wolves and dogs to have occurred. Other scholars believe it happened between 12,000 and 20,000 years ago, in the Upper Paleolithic period. Both humans and wolves inhabited much of the

At last count, the family Canidae (derived from the Latin word *Canis*, meaning dog) was made up of thirty-five different species organized into fourteen different genera. The most familiar genus, *Canis*, includes various species of wolves, jackals, coyotes, and dingoes, as well as the domestic dog.

Foxes are part of the Canidae family (popularly known as the dog family) even though many of them have more in common behaviorally and physically with felines than canines. They make up the genera *Dusicyon* and *Vulpes*. Many more exotic canids boast their own genera. For example, South America's crab-eating fox is a *Cerdocyon* and the nocturnal raccoon dog of Southeast Asia is a *Nyctereutes*. Despite popular belief, the hyena is not a member of the Canidae family. Genetically it is closer to felines and actually consists of four different species within its own family, Hyaenidae.

Taxonomy, or classification of animals, is not a static science. Technological advances in genetics mean that changes to Carolus Linnaeus's system of taxonomy are inevitable. Once, for example, the Ethiopian, or Simien, jackal from the Bale Mountains of Northern Ethiopia was classified as a kind of fox. But recent DNA testing caused scientists to change their minds—and the animal's name. It's now known as the Ethiopian wolf.

earth's surface by that time, but it's still unclear where exactly the domestic dog first emerged—indeed, this process probably occurred in several different areas. Evidence suggests the Middle East and Southwest Asia as the major centers of domestication. Fossil bones of modified wolves at least 14,000 years old have been unearthed in various locations there. These were most likely descendants of the Indian Plains wolf, a long-extinct lupine thought to be the earliest ancestor of a great number of modern dog breeds.

What led humans to raise wolves? It could have been something as simple—and compelling—as companionship. Paleolithic humans may have taken wolf cubs, perhaps the orphans of the wolves they killed, from their dens and raised them. These semi-tame wolves would likely have continued to live and breed in and around human settlements, gradually producing slightly tamer generations that became more and more comfortable around humans over time. Particularly vocal wolves may have been mated to produce offspring who would warn of intruders in the night, creating the first guard dogs. At some point, semi-domesticated wolves began accompanying humans on hunting expeditions. Before too long, man must have deduced that active selective breeding would ensure this mutually beneficial relationship. Later, superb hunters may have been mated to

Wolves typically howl as a means of vocal communication. They also bark, but do so much less frequently and at a lower volume than domestic dogs, and usually only at close quarters. Wolves that barked more than others may have been singled out and mated to produce the first guard dogs for early human encampments.

others exhibiting the same skills to produce highly efficient trackers. The transformation from wolf to domestic dog was under way.

DOMESTICATION AND ITS EFFECTS

As the wolf was modified, it became a smaller animal, a typical phenomenon during the domestication process and the result, many archeologists and anthropologists believe, of heterochrony: a change in hormones and hormone levels as the animals become dependent on humans. Other scientists point to a more active human role in the breeding process. Some traits, most likely small size and juvenile characteristics, were considered desirable. Others appeared as inadvertent consequences of trying to perpetuate favorable traits.

Fossil evidence suggests that it took only a few thousand years for dogs to develop the plethora of sizes, shapes, and colors evident today. Two skeletons found in Idaho's Beaverhead Mountains and thought to be around 9,000 years old show one medium-sized, retriever-type dog and a smaller, beaglelike animal. Three-thousand-year-old mummified remains found in Arizona in 1905 demonstrate that the evolution of the dog had taken another leap forward in the intervening years. They reveal a small, terrier-type dog with a long black and white coat, the first evidence of a departure from the wolf's short gray coat.

The incredible diversity within *Canis lupus* also provided rich material for selective breeding. For example, the Canadian timber wolf can weigh as much as 175 pounds, while the extinct Japanese wolf, the likely precursor of some Eastern breeds, was a featherweight at no more than thirty pounds. Such variation could easily have been exploited to create a breed of smaller dogs. Similarly, traits such as hair length, color, ear size and even behavior could have been nurtured by mating dogs with similar qualities.

A WOLF IN DOG'S CLOTHING

We may have dressed domestic dogs in a wide array of colors and patterns, and molded them into a multitude of shapes and sizes, but the nature of the wolf lives on under an exterior coat of domestication. The only physical or behavioral differences between wolves and dogs were created by humans. Wolves (*Canis lupus*) and domestic dogs (*Canis familiaris*) have been regarded as two different species within the genus *Canis* since Swedish biologist Carolus Linnaeus created a system for classifying animals in 1758. Scientists now claim that the separation is an artificial one. In 1993, both the Smithsonian Institute and the American Society of Mammologists reclassified the domestic dog as *Canis lupus familiaris*, a sub-species of the gray wolf. Genetically, even after thousands of years of domestication, wolves and dogs have almost identical DNA patterns. They can also interbreed under natural conditions—one of the defining characteristics of members of the same species.

RETURN OF THE WOLF

Once the farthest ranging of all mammals except humans, the wolf has been hunted close to extinction. Today it lives only in isolated pockets in northern forests, mostly in Canada and Alaska. That may soon change. Recently, various government-sponsored recovery programs have reintroduced wolves to areas where they once thrived.

The red wolf (*Canis rufus*), considered by many to be merely a smaller and differently colored sub-species of gray wolf, once roamed much of the southeastern United States. Extinct in the wild for several decades, it was successfully reintroduced to parts of its former habitat in 1988.

In the U.S. Northwest and Midwest, similar programs to return the gray wolf to its former range have enjoyed success, but not without resistance from ranchers: They claim that wolves are a predatory menace to their livestock and their domestic dogs. The most notable of these programs recently released ninety gray wolves into Yellowstone Park, where they'd been absent since 1926. The success of such projects has led New York State wildlife authorities to consider a similar effort with the eastern timber wolf in parts of Adirondack Park.

Wolves destined for Yellowstone National Park are moved into holding pens.

DOG
·DESIGN·

Despite their differences in size and shape, from the tiniest desert-dwelling fox to the largest of the northern wolves and giant domestic dog breeds, canids almost invariably share at least one quality: They are strong, lean running machines. And with good reason. To survive and thrive throughout the ages, canids had to be swift and strong to catch and take down their prey, and needed powerful, unyielding jaws and flesh-shearing teeth to consume it. And they had to do it in a wide range of environments that began with their original grassland home and eventually spread to the utmost ends of the earth.

Today, almost all canids have lithe bodies with proportionately long legs. The only exception in the wild is the short-legged, somewhat otterlike bush dog, a strong swimmer from Central and South America that preys predominantly on large water rodents, and the equally short-limbed raccoon dog of Asia, which looks more like—naturally—a raccoon.

In domestic dogs, selective breeding for a variety of traits that occur naturally has widened the normal range of canine build. Short-legged miniature and dwarf dogs, who never would have survived in the wild, have prospered because of human intervention. On the other end of the size scale, the coupling of very large dogs eventually resulted in the giant breeds such as mastiffs and the Irish wolfhound. These dogs are taller, heavier, and more powerful than all but the largest of wolves. But despite the vast physical differences displayed among domestic dog breeds and other canid species, they all share many anatomical features.

WHAT'S BRED IN THE BONE...

Size differences aside, all canids possess virtually the same skeletal structure underneath their usually furry exterior. A sturdy skull with deep sockets to house the eyes and inner ears surrounds and protects the animal's brain. As in other carnivores, seven cervical vertebrae attach the skull to a spinal column composed of thirteen or fourteen thoracic vertebrae and six to eight lumbar vertebrae.

Like humans, and unlike cats, canids trade some spinal flexibility for additional strength and skeletal rigidity. Bony extensions, called dorsal spines, on all of the spinal vertebrae act as attachment points for the strong muscles that provide powerful backs and necks, useful in many wild canids for holding prey with their mouths. Discs of tough joint cartilage wedged

Today's miniature or toy dog breeds are the product of both nature and man. Two forms of size reduction occur naturally: miniaturization, a proportional reduction in the size of all the bones in a dog's body (Yorkshire terriers and Chihuahuas are examples), and dwarfism (displayed in the dachshund above), a shortening of the limbs on a normal-sized body. Humans then exaggerate these differences when they breed for these traits.

Opposite: A testament to their hunting nature, most canids—including wolves, coyotes, and many domestic dogs—possess long legs and deep chests for running stamina, as well as powerful jaws and sharp teeth for killing and eating prey.

RESHAPING THE WOLF'S SKULL

 One of the few ways in which domestic dogs have changed dramatically from their wolf ancestors is seen in the shape of their skulls. Domestication and the subsequent breeding for specific, desirable head shapes have resulted in three basic skull types.

Dogs bred for exceptional sight, such as Salukis and greyhounds, sport slightly elongated versions of the wolf's cranium. Called "dolichocephalic," this skull shape features a long and relatively narrow snout.

In many scent hounds and pointers, the wolf's skull has been reduced in length. These "mesocephalic" skulls feature somewhat broader snouts and larger nasal chambers that help improve scenting abilities.

Boxers, bulldogs, and some other so-called "fighting" dogs possess a much shorter snout. Called "brachycephalic," this compact skull type is thought to result in more powerful jaws—and in extreme versions, such as the pug and the bulldog—a variety of health problems. Along with sometimes having painfully overcrowded teeth, dogs with reduced snouts cannot cool themselves properly by evaporating water through their nasal cavities, as most canids do. These pets are at particular risk of overheating in warm weather.

Dolichocephalic skull (Greyhound)

Mesocephalic skull (German short-haired pointer)

Brachycephalic skull (Bulldog)

snugly between each of the vertebrae act as shock absorbers. Together with the dorsal spines and back muscles, the taut configuration of the vertebrae and joints maximizes a canid's ability to run and jump. As a result, the members of the dog family are not as flexible as cats, whose vertebrae can twist in a variety of directions, permitting a lateral motion that would hamper a canid's speed. Still, canids have relatively elastic backs, flexing and extending their trunks with far greater ease than another quick, four-legged animal, the horse.

Between fourteen and twenty-three coccygeal vertebrae complete the spinal column and make up the canid's tail bone. Wolves, African wild dogs, and such domestic breeds as greyhounds, who run extensively, have longer tails, which they use to provide additional balance during sharp turns at top speed. Although a few domestic dog breeds do have naturally short tails, such a feature is more often a result of the controversial practice known as docking: The tail is cut off at one of the coccygeal vertebra when the dog is just a few days old. Back when most breeds were working dogs, docking was done to prevent injury in the line of duty. Today the practice is much harder to justify. It is carried out in the name of tradition and to comply with the longstanding breed "standards" for appearance dictated by kennel clubs *(pages 173 to 175)*. Tail docking is now illegal in many countries, including Great Britain and Australia.

Attached to the thoracic vertebrae, a canid's ribs run down each side of the body. Some are secured to the sternum, or breastbone; the last pairs are floating ribs that are attached only to the thoracic vertebrae. Together, the ribs form a protective cage for such vital organs as the heart and lungs. The sternum itself is flexible, moving in and out as the animal breathes. The length of the ribs dictates the size of the canid's chest, and canids who run a lot need a deep one. To meet their demanding oxygen requirements, wolves, African wild dogs, and long and lean domestic dogs like borzois and greyhounds have especially large chests to accommodate a big heart and high-capacity lungs.

TAKING THINGS IN STRIDE

From the position of its legs to the structure of its bones, the canid's skeletal makeup is marvelously adapted to providing a long and efficient stride.

The canid's rigid skeletal structure may limit its ability to twist and turn, but powerful back muscles and long limbs make specimens such as this German shepherd dog explosive runners.

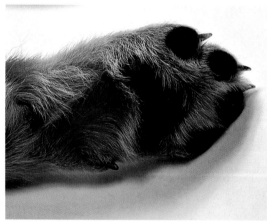

Four toe pads and claws are located below the toes on the underside of a canid's paw. They and a large common pad in the center act as shock absorbers. A smaller stopper pad, located higher up on the back of the leg, provides a better grip for walking or running on slick surfaces.

The scapula, or shoulder blades, to which the long forelimbs are connected through a shoulder joint, are not attached directly to the vertebrae or ribs. Instead, these flat, rectangular bones are held tightly in place by muscle alone along the sides of the body, permitting a longer stride. Like the cat, the canid's miniscule collarbone, or clavicle, has retreated deep inside a thick chest muscle. This allows the limbs to be brought farther under the body, making their forward-backward motion more efficient. The animal's long hind legs, attached to the lumbar region of the spinal column through the hip joints and a flat and elongated pelvis, provide the propulsive thrust needed for running and jumping. The impact of their powerful strides is absorbed by the limbs. For added stability, the radius and ulna bones that make up the lower forelegs are bound tightly together with ligaments, and the scaphoid and lunar bones of the wrist are permanently fused.

Every little bit counts when it comes to canid locomotion. All members of the dog family move in a digitigrade fashion: walking and running on their toes. This lengthens their stride ever so slightly, while improving their traction. But long legs and a graceful stride don't always make for superior running ability. For example, the maned wolf has exceptionally long limbs, but despite appearances the animal is not a particularly fast runner. Instead, scientists believe that the wolf's long legs serve primarily to help the animal see over the tall vegetation of its South American grassland habitat.

PADS, PAWS, AND CLAWS

On the underside of the dog's paws, several thick, leathery pads serve as protective shock absorbers to cushion landings and provide a good grip when running or walking. Sweat glands in the pads help keep them soft and produce a scent dogs use when marking objects (*page 52*). If cut, the paw pads will bleed profusely.

Almost all canids have four regular toes and blunt, non-retractable claws to help grip the ground while they run. And all canids except for the African wild dog have a fifth vestigial toe and claw, called the dew claw, on the forepaws, positioned much like a thumb. Domestic dogs and dingoes also have dew claws on their hindpaws. Some domestic breeds—the Great Pyrenees, the Briard, and its relative, the Beauceron, for example—have double dew claws on their hindpaws, although nobody is sure why. While dew claws can be used to restrain prey, they may be leftovers from a time when dogs climbed trees.

Because they are not dulled or worn down by walking, dew claws are quite sharp. With domestic dogs, they need to be trimmed more often than the other claws to prevent them from getting caught and torn on objects.

Opposite: Superb athletes, wolves can propel themselves over the landscape and through the air, thanks to powerful hind legs and strong back muscles. The rigidly designed skeleton that makes canids such accomplished runners, jumpers, and even swimmers does, however, limit their ability to climb. Of all the members of the canid family, only the agile gray or tree-climbing fox of the Americas can climb well, thanks to a more flexible skeletal make-up and limbs that rotate freely.

CANINE THERMOSTAT

Like all mammals, the canid family is able to control body temperature through various physical functions. Called thermoregulation, this ability allows canids to thrive in all types of climates, from the sub-arctic to the Sahara.

Most mammals sweat to keep cool. But an absence of sweat glands on the canid's skin forces it to rely on other means. In temperate weather, canids simply breathe in through their noses to stay cool: Nasal glands secrete fluid, and its evaporation within their nasal chambers creates a cooling effect. When the animals become significantly warmer, either through a change in air temperature or through physical exertion, they secrete even more fluid than usual. To cool down, they begin to pant with open mouths to breathe in more oxygen and

help speed up the evaporation and cooling process. The lolling tongue increases the surface area where evaporation can take place.

Canids also rely on their coats to control body temperature. Species such as wolves who inhabit cold climates sport double coats, with a thick, insulating under-layer to retain heat and keep the cold at bay. The fur can be puffed up to trap warm air in cold weather or held closely against the skin to release heat on warm days.

Even the size of a canid's ears plays a part in temperature control. Such hot-weather species as the fennec fox, a tiny African desert dweller, sports oversized ears to maximize heat loss. The arctic fox, on the other hand, has miniscule ears to minimize the effects of the frigid weather in its northern homeland.

A dog's pant isn't just a sign that he's tired. It serves to evaporate the liquid in his mouth and on his tongue, cooling the animal down.

Right: Wolves have exceptionally thick, two-layered coats to protect them from extreme weather. Typically they sport solid gray, brown, or black coats, composed of light to dark color-banded hairs, called agouti, like the gray wolves shown at right. But variations in certain sub-species of the wolf are thought to be responsible for the wide variety of coat patterns and colors found in domestic dogs today. The arctic wolf of the tundra is often pale gray or all white, while Asian and desert-dwelling wolves have sandy-red coats.

Above: The white winter coat of the arctic fox serves as camouflage in its snowy northern environment. Far less common, a blue variety can be found in less snowy coastal regions. In both types, a long, weatherproof topcoat and soft, thick undercoat can keep the fox warm in temperatures as low as -75°F (-59°C). In the spring, the arctic fox sheds its heavy undercoat to cope with temperatures as high as 70°F (21°C). Once the snow begins to disappear, the fox's coat turns from white to brown to match the mottled habitat. The blue variety turns to a dark brown or blue-gray in the summer.

Sometimes they will even curl and grow directly into the pad below, causing great pain. Dew claws are commonly removed early in life to eliminate these problems. The breed standards for some purebred dogs actually require their removal, a practice denounced as cruel and unnecessary by some dog lovers; others consider it to be a preventive measure in the dog's best interest.

Unlike cats, very few dogs possess an extra toe, a trait known as polydactylism. One exception is the lundehund or Norwegian puffin dog, which may have six toes on each foot, as well as double dew claws. The extra digits enabled the breed to negotiate the rocky cliffs they once climbed to retrieve nesting puffins, which used to be a traditional delicacy for Norwegians.

HAIR OF THE DOG

There are almost as many types of domestic dog coats as there are dogs, with various lengths, textures, types of hair, and proportions combining to create an array that encompasses everything from the short, fine coats of pointers to the long, thick, bristly hair of Samoyeds. A dog's coat is generally made up of primary and secondary hairs. The long, rigid, and coarse primary hairs, also called guard hairs, form a protective, water-resistant topcoat in many wild canids and domestic breeds. Even in rain and snow, this coat keeps the animal dry underneath. The shorter and softer secondary hairs provide dogs with a thick, soft underlayer that insulates the animal like a good down coat.

Density of hairs is another variable in the coats of dogs. Most terriers have relatively thinner coats, while cold-weather spitz-type dogs such as Alaskan Malamutes and Chow Chows have thick, dense coats, sometimes with more than 3,900 hairs per square inch (604 per square centimeter). But no matter how dense a dog's fur is, it is the proportion of hair types that affects how soft or harsh its coat will be. A coat containing a high number of primary hairs will most certainly be medium to long in length and rather coarse, while one with a higher concentration of down hairs will be short and soft.

UNUSUAL COATS

The uneven proportion of primary and secondary hairs sometimes accounts for what is known as "wirehaired" dogs. Made up almost exclusively of densely packed primary hairs, this coarse and wiry coat appears in many breeds, from small terriers and dachshunds to large dogs such as the wirehaired pointing griffon. The coarse outer layer of wirehaired coats offers the dogs who wear them excellent protection from the elements and rough underbrush—and even the bites of other animals.

Mother Nature offers up a variety of other unusual coats in domestic dogs. Genetic changes, later accentuated through selective breeding, have resulted in both hairless and long-haired dogs, neither of which would have survived natural selection in the wild. Hairless dogs such as the Mexican and Peruvian hairless and the Chinese crested dog *(right, top)* sport tufts or wisps of hair on their heads but none on their torsos. They are unsuited to cold climates and must even be protected from direct sunlight to avoid burns.

Dogs with exaggerated hair length are well suited to cold weather, but such long-haired breeds as Afghan hounds and Old English sheepdogs require constant grooming to avoid hair matting. The komondor, a sheep herder from Hungary, may sport the most unusual coat of all: Its long hair is naturally braided into felt-like cords.

Hair growth in dogs is normally a cyclical process, with periods of active hair growth—most protracted in long-haired dogs—followed by resting periods. Most dogs shed and then replace their "winter coat" in spring, the result of glandular and hormonal effects, as well as changes in daylight hours and temperature. As hairs fall out, new ones grow in to take their place. After a growth and resting phase in the summer, shedding resumes in the fall as thicker down replaces summer's light coat, readying the dog for the cooler winter months.

Not all dogs follow a seasonal shedding cycle. Some shed at a consistent rate throughout the year, especially those who rarely venture outdoors. And some domestic dogs—poodles, for instance—hardly shed at all. They require frequent clipping, however, to take care of what nature neglects.

A range of coats: Hairlessness, as seen in the Chinese crested dog *(top)*, was the result of a genetic accident, later perpetuated by selective breeding. The wirehaired coat of the dachshund *(middle)* features a higher quantity and density of the harsh primary hairs in the dog's coat. The long hair of the Lhasa Apso *(bottom)* is due to a naturally prolonged growth phase that was emphasized by selective breeding for thermal protection and aesthetic reasons.

THE
·SENSES·

From remote northern forests to crowded urban dog runs, canids rely on smell, taste, hearing, vision, and touch to reach out and collect information about the world around them. Just which senses are most acute depends on their lifestyle and means of survival. Nocturnal hunters like foxes possess extra-keen hearing and an ability to see in low light that enable them to track the small, foraging rodents that make up a portion of their diet. But their sense of smell is not nearly as sharp as that of wolves, who must track much of their prey over long distances and often through thickly wooded areas. The same principle of selective adaptation and enhancement of the senses holds true for the various breeds of domestic dogs. Scenthounds, such as beagles and bloodhounds, can follow even the faintest of scent trails through dense underbrush and urban alleyways. Their sense of hearing, however, is nothing more than ordinary.

LED BY THE NOSE

Smell, or olfaction, is the canid's most powerful and sophisticated tool for collecting information about his physical world, as well as the emotional state of his animal and human neighbors. With scent receptors lining the inside of the nasal cavity and the roof of the mouth, a canid can fully enjoy the glory of a fresh kill, analyze another's emotional or sexual status and perhaps even sense the mood of a human companion.

Observe any domestic dog's greeting rituals, and you'll see that no other sense predominates more than smell. In the brief time that it takes dogs to sniff one another, they will have discerned the other's age, sexual status, social ranking, and even, to some extent, recent meetings and activities. Humans are essentially greeted in the same way, and for much the same reason. Dogs recognize their human family members not only from sight and sound, but also from our scent, often layered with information such as our recent meals.

Canids detect and analyze smells much as we do: The air that is breathed in or sniffed reaches scent receptors in the nose and mouth. Then the information gathered by the receptors is relayed by the olfactory nerve to the brain. The canids' advantage lies in their ability to take in more information at a time, and better distinguish the range of scents with every breath.

The slits on the sides of a canid's nostrils allow them to flare to a greater extent than humans' round nostrils. With more air inhaled per breath, the

A lot of information can be garnered from a quick sniff. Domestic dogs glean both physical and emotional details about one another through their olfactory senses. In addition, when dogs of similar stature meet up, one often bumps the other with the rear of its torso to get a feel for its potential rival's body mass.

Opposite: This coyote appears to be tasting the air, and in a way, it is. In what may seem to observers to be a hybrid of grin and grimace, coyotes, jackals, and most other canids use a so-called flehmen response to draw air into the roof of the mouth, where the vomeronasal, or Jacobsen's, organ lies. This bundle of olfactory nerve endings allows the canid to detect and analyze what are usually sex-related scents present in the air.

NOSY BARKERS

The tracking abilities and achievements of scenthounds such as the famous bloodhound are well documented. These nasally gifted canines can doggedly follow scent trails to find everything from lost children to escaped prisoners. But special scenting abilities are not the sole domain of scenthounds. Collies, German shepherd dogs, schnauzers, and other terriers possess exceptional scenting skills. This ability, combined with their easy trainability, accounts for their use by law enforcement agencies to sniff out drugs and bombs.

Recently, medical science began putting dogs' incredible sense of smell to good use. Special cancer-detecting dogs are capable of locating traces of skin cancer on human beings. Given a sample whiff of a scent, be it the dead skin cells present in a piece of a lost child's clothing or cancerous cells in a test tube, special scent dogs can locate the exact replicas of the scents in question with incredible accuracy.

Although the best explanations for these amazing talents lie somewhere in the mix of good genes and superb training, other factors influence the dogs' scenting abilities. Dogs are known to track more effectively if they are hungry, and male dogs—although nobody is quite sure why—make for better tracking dogs than their female counterparts.

Originally bred to track the elusive hare, the harrier is known for its scenting ability.

canid's nose has access to more scent molecules at a time. And with its malleable nostrils, it can also more easily detect the direction from which the scent originates. Long before it can see prey, the canid knows where to find it. In addition, the moisture on the outside of its nose helps capture airborne scent molecules. Canids lick their noses often to keep them moist for this very purpose.

The scent molecules essentially "stick" to the canid's nose, dissolve in the moistness of saliva and nasal secretions, and then move into the nasal chamber with each sniff. Once scent molecules are inside the nasal cavity, they settle over a complex network of membranes and thin bones, called turbinates, which contain sensory cells or receptors. The ensuing chemical signal is then passed along as a neural response to the brain's olfactory bulb for interpretation. The incredible scrolling of turbinate bones, and their resulting high-surface area for capturing and directing odors toward receptors, contribute to a canid's uncanny ability to distinguish smells. By comparison, a human nose has only three little curled bones, with their membranes' relatively smaller surface areas, to capture and analyze smells.

NOSE POWER

Any way you look at it, members of the dog family are more physically suited for smelling than humans. For starters, their two hundred and twenty million scent receptors outnumber ours by more than forty to one. An extremely large part of the canid brain is devoted to smell alone. Canids "remember" previously analyzed scents for future reference. In fact, the propensity for some scent recognition is passed down along generations in much the same way as other inherited traits and is specific to a dog's means of survival and lifestyle. For example, beagles were bred originally as hunting hounds, so they pick up on the scent of a fox infinitely more quickly than will a bulldog, whose original purpose was to bait and fight bulls. In addition, hunting canids' superior ability to distinguish among closely related smells helps them to focus on the source of the scent without being distracted or otherwise losing the trail to its next meal.

The larger the nasal chambers, the greater the number of receptors—and the better the tracker. Bloodhounds and other scent hounds, the super smellers of the dog world, have bigger nasal chambers—and a finer sense of smell—than those top-notch trackers of the wilderness, wolves. These breeds of domestic dogs also use other features to help get the job done. Huge drooping ears stir up air currents along the ground and large, pendulous, moist lips help to capture even more scent molecules.

Canids employ another, lesser-known weapon in their olfactory arsenal. A pouch-like organ on the roof of the mouth of some mammals, called the vomeronasal, or Jacobsen's, organ, is thought to capture and analyze pheromones, hormonally produced body scents that convey information about the sender's sexual status. Coyotes and jackals employ an odd combination of grin, gape, and grimace known as the flehmen response to draw scents into this pouch. So, too, do goats, and both wild and domestic cats. Although airborne odor molecules are not as easily brought to the vomeronasal organ without the flehmen response, male domestic dogs and wolves, who do not employ the technique, still use the vomeronasal organ to detect females in heat. Some females of these species are also thought to utilize it on occasion.

TONGUES AND TASTE

Smell initially attracts a canid to food, but once its attention has been captured by an attractive scent, the animal's tongue and taste buds take over. Not nearly as refined as the canid's sense of smell, tasting ability works in a decidedly broader fashion. While humans have more than 9,000 taste buds or receptors on the tongue, canids have only about 2,000 to help them

CANINE TEETH

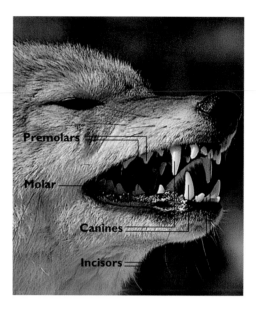

Premolars

Molar

Canines

Incisors

This snarling wolf reveals its deadly array of teeth. As omnivores, canids need a variety of specialized teeth to eat the different types of foods in their varied diets. Canids, including domestic dogs, have forty-two teeth, although there are a few exceptions: the bat-eared fox has forty-eight, dholes forty, and bush dogs thirty-eight. Two rows of six small, curved incisors help the dog maintain a firm grip on prey, while four large, pointed canine teeth tear at it. Sixteen sharp premolars, including the carnassials, allow the dog to cut and slice flesh while ten powerful, grinding molars mean the dog can chew the non-flesh parts of its diet—typically vegetables, other vegetation, and grains.

Dogs use their large, spongy tongues like ladles to lap up water. Special taste receptors on their tongues are believed to allow canids to recognize chemical composition variances—such as salt and impurities—in different water sources.

A dingo pricks up its ears to locate a sound before funneling it into the inner ear for transmission to the brain. Many canids can even scan their environment for sounds using one ear at a time.

judge flavor. Although canids, like human beings, have taste bud receptors on their tongues that register sweet, sour, salty, and bitter flavors, their general reactions to food can be divided into three categories: good, passable, and bad. Receptors for sour and bitter foods are spread over the surface of the tongue, while those responsible for detecting saltiness are located along the sides and at the back. Sweet tastes, to which canids are more or less indifferent, are picked up by receptors at the tip and along the edges of the tongue. Texture is also thought to play an important role in the canid's dietary habits, and pressure and texture receptors on the tongue's surface add to food's sensory pleasure.

What canids may be able detect that humans can't are taste differences in water. Extra receptors for this purpose are believed to allow both dogs and cats to detect subtle variations in the chemical, salt, and other mineral contents present in water.

Like humans, canids are omnivorous. Despite displaying preferences for meat over other foods, and a fondness for beef in particular, canids will also eat cereal, fruit and vegetables. No matter how good food looks, though, the animals' sense of smell will generally warn them if it is spoiled. But if their sense of smell is impaired due to illness and they eat bad food, the vomiting reflex allows them to regurgitate the offending morsels before any real damage can be done.

LEND ME YOUR EARS

In the vast expanse of the frozen tundra, a lone wolf perks up its head, twitches one ear, cocks its face to one side, and then sets off in hot pursuit of distant prey. All canids have the ability to detect and distinguish sound at a great distance, further localize it by moving their ears, and then zero in on its source. Even domestic dogs enjoy these sensory vestiges of their wild cousins, something that is abundantly evident to dog owners who watch their pets react to the sound of a can opener in action in another room.

Although their hearing capacity is nearly identical to that of humans at low frequencies, canids can hear a higher and broader range of frequencies. Sound frequency is measured in Hertz (Hz), or the number of sound waves that pass one point in one second. The higher the pitch, the higher the frequency, and the more sound waves that move past that point per second. While humans can generally hear frequencies between one and 20,000 Hz, canids stretch that range upward to approximately 40,000 Hz, a sensitivity that allows them to detect the high-pitched squeaks of burrowing rodents and other prey. Hearing range differences do occur among wild species and even among domestic breeds. Coyotes are thought to have the most sensitive hearing of all canids, with a range of up to 80,000 Hz. In domestic dogs, greyhounds have been known to detect sound frequencies of more than 60,000 Hz.

While large ears are valuable heat-loss tools for some desert-dwelling canids and others hailing from hot climes, proportionately big ears serve other purposes. The ones that adorn the bat-eared fox shown at left help funnel the high-pitched sounds of insects and small rodents into the inner ear.

Canids also can discriminate between nearly identical sounds. With a hearing range of eight-and-a-half octaves (the same as humans), they can recognize differences in pitch of as little as one-eighth of a tone. A domestic dog can thus prepare to appropriately welcome its owner at the familiar sound of his shoes as they scuff the front walk. A less friendly greeting may be in store for strangers whose footfall sounds the slightest bit different to the dog's ears.

Thanks to a pair of highly mobile outer ears composed of flexible cartilage, canids can scan their environment for sounds. Studies have shown that members of the dog family can locate the source of a sound in six one-hundredths of a second, sometimes with a single ear doing the scanning. Sound waves move from the outer ear through the middle ear, where they are amplified by three tiny bones—the malleus, incus, and stapes (commonly called the hammer, anvil, and stirrup due to their shape)—and then are transmitted to the cochlea, part of the inner ear. Here they are converted to chemical signals that are relayed to the brain.

As in humans, the canid's ear also contains a mechanism that helps maintain balance. In addition to the malleus, incus, and stapes, a pair of semicircular canals in the inner ear—the saccule and utricle—serve as the organs of balance. The saccule and utricle are lined with tiny, sensitive hairs and are filled with fluid and crystals, all of which help detect the slightest change in the position of the dog's head. This information is sent to the brain, which then sends a signal back, enabling the animal to fine-tune the orientation of its head.

HEARING FOR HUMANS

The dog's acute sense of hearing and considerable intelligence help it provide an invaluable service to human beings. Like guide or "seeing-eye" dogs for the blind, special signal or hearing dogs are trained to alert hearing-impaired people to the sounds of everyday life. A number of charitable organizations across the world offer these special dogs free of charge to an ever-growing number of applicants.

Hearing dogs are usually young, mixed-breed dogs rescued from animal shelters, Before entering service they must go through an extensive training period, which consists of teaching them to alert their masters to everything from smoke alarms, timers, doorbells, telephones, and alarm clocks to crying babies, wailing sirens, and even the call of their name. In most cases, the dogs are trained to touch or wake up their owners, then run to the source of the sound. Hearing dogs are also taught to respond to their masters' commands. They learn to heed calls to sit, stay, and fetch through special hand signals.

Hearing dogs do more than instill confidence and a feeling of security in their partners. They also provide loving companionship—a gift that many hearing-dog owners consider to be as important as their pet's role as surrogate listener.

DOMESTIC DOG EAR SHAPES

The ears of wild canids come in only one shape: natural. But among domestic dogs, selective breeding has produced a number of different ear types, each with its own name, courtesy of dog breeders, fanciers, and kennel clubs. Among the most common varieties are the erect or "pricked" ears sported by many of the more wolf-like breeds, such as German shepherds and basenjis, and by semi-domestic canids, such as dingoes and New Guinea singing dogs. Most hound dogs feature hanging or "pendant" ears, while many terriers have semi-erect "button" ears in which the tips fold over and hang in a V-shape, partially covering the opening. The "rose" ears found in some bulldogs and pugs, a variation of a drop ear, is folded inward along the back edge, with the tip curving over and back to show the opening to the ear.

**Pricked ears
(Basenji)**

**Pendant ears
(Basset hound)**

**Button ears
(Airedale hound)**

**Rose ears
(Pug)**

THE EYES HAVE IT

While it may be tempting to compare canid vision with human vision, it's very much the proverbial case of judging an apple next to an orange. Canids have far different visual needs than we do, and have adapted different visual abilities because of these requirements. Humans have superior depth perception, and can see colors and focus on objects better than canids. But canids possess exceptional lateral and distance vision, detect motion better, and can see well in poor light—adaptations that no doubt contribute to their hunting success.

The key to vision lies in the eye's ability to collect and then use light efficiently. Two types of photoreceptor cells in the retina help gather light: rods and cones. Rods do most of their work in dim light, while cones are responsible for color vision and only work in bright light. A retina composed mainly of cones evolved in primates, including humans, allowing us to differentiate between colors and to see well during the day. However, since canids are mostly diurnal, with the occasional nocturnal hunting need, their visual system has evolved to function under conditions of limited light. In fact, their vision is actually sharpest at twilight, the result of the high percentage of rods on their retinas.

A reflective layer of cells behind the retina helps enhance the canid's night vision. Called the tapetum lucidum, this layer of cells re-stimulates light-collecting rods to maximize the use of available light. More famous for producing the eerie glow-in-the-dark eyes of cats, the tapetum lucidum is nevertheless present in canids, as dog owners can attest who have seen a car headlight strike their pets' eyes at night.

Eye placement varies greatly among the domestic dog breeds and has a direct impact on the dog's visual abilities. Sighthounds like these Salukis have frontally placed eyes that provide them with the superior binocular vision, or depth perception, needed to hunt prey. But there's a trade-off: the dog has limited peripheral, or lateral, vision.

The relative absence of cones in the retina means that canids live in a mostly black-and-white world. However, tests have shown that the type of cones present allow canids to see some hues of blue and yellow. They cannot distinguish green, red, orange, or yellow-green, and probably see these colors as shades of white or gray. However, canids do have the upper hand on humans when it comes to real grays, detecting more subtle shade variations than we do.

The low number of cones adversely affects canid vision in another way. Canids lack what is known as a fovea, an area at the center of the retina with a highly concentrated mass of receptor cells. The fovea is responsible for the visual acuity of humans—our eye for detail. Instead, canid eyes have a long, horizontal "visual streak" made up of a high concentration of rod cells. Because of this, the visual acuity of canids is an estimated six times poorer than ours. Using the Snellen fraction, the standard indicator of human resolving power, it is estimated that the typical domestic dog's vision is 20/75. This means that a dog must be twenty feet from an object to make out the detail that people with good vision can see from seventy-five feet.

Despite their lack of good color perception and limited ability to distinguish detail, canids have a superior ability to detect movement and see at great distances—important factors in their hunting prowess. Both of these attributes, largely due to the abundance of rods in the retina, are well-suited for hunters on open grasslands, such as the visually gifted African wild dog, and for domestic herders like sheepdogs. Some domestic dogs that can detect a stationary object from about 650 yards (594 meters) away, will be capable of spotting the same object, when moving, at 875 to 985 yards (800 to 901 meters). Sheepdogs are thought to detect motion, be it the signals of their shepherd partners or the stealthy approach of an encroaching predator, from almost 1,100 yards (or one kilometer) away.

The position of the eyes plays an important role in both lateral vision and depth perception. Wolves and some domestic dogs such as German shepherds have eyes set far apart, providing them with excellent peripheral vision. Skull and nose shapes and eye placement and angle vary greatly from breed to breed, but on average, the field of vision for domestic dogs is 240 or 250 degrees out of a possible 360. Humans, by comparison, are limited to 180 degrees. Sighthounds such as Salukis, borzois, and greyhounds have frontally placed eyes, which results in a sacrifice of lateral vision for better depth perception. This so-called binocular vision is the result of the overlap of each eye's field of vision. Binocular vision provides the depth perception necessary for predators to time their leaps in the quest for elusive prey. This ability varies greatly among domestic dogs, ranging from thirty to sixty degrees—far less than the 140-degree vision possessed by humans.

TOUCHY, TOUCHY

Neglected by physical science because of its relative simplicity, the canine sense of touch is an undeniably important factor in a dog's emotional well-being. One of the first senses to develop in a newborn, the sense of touch soothes newborn puppies and cubs and serves as a crucial tool in a growing canid's upbringing. Deprived of touch during the vital socialization period *(page 103)*, a domestic dog will grow up wary of humans and perhaps other dogs. The kind touch of a human companion remains the greatest reward that can be bestowed upon a dog.

To reach out and touch their surroundings, canids count on highly sensitive whiskers, or vibrissae, on their muzzles, jaws, and over their eyes. Embedded in sensitive nerve- and blood-rich parts of the skin, the canid's whiskers feed it information on air flow, detecting even the slightest change. When the whiskers come in contact with an object, they can detect both its shape and texture. Nerve endings also cover the canid's entire body; the feet and paws are by far its most sensitive areas.

While they may be less pronounced than on a cat, the whiskers of a dog are nevertheless an important sensory tool. Wild canids such as the short-legged bush dog who sniff frequently at the ground for information have well-developed whiskers on the underside of their jaws.

THE LIFE
· STAGES ·

Newborn puppies sleep almost constantly for the first week or two of their lives. Born blind and deaf, this two-day-old Labrador retriever puppy is still capable of feeling, smelling, and tasting. By two or three weeks of age, it will start seeing and hearing, while its other senses continue to develop. And by the end of the first month, it will begin to growl, howl, and bark.

The life of a dog lasts but a small part of our own. Aside from the enriching companionship pet ownership provides, living with a dog from puppyhood through old age provides us with a unique opportunity to observe and participate in the entire birth-to-death cycle of life.

Born blind and deaf, and unable to control their own body temperature, both wild and domestic puppies rely on their mothers for everything in the days following their birth, especially nourishment and warmth. For the first 24 hours, the mother will not leave her pups. Over their first few days, newborns nurse every few hours. The mother's nutrient-rich colostrum quickly turns to a high-fat milk that helps keep them warm and well nourished. Barely able to even crawl, the young tread on their mother's teats with their paws to stimulate milk flow, and each pup finds a nipple to feed on through smell and special heat-seeking sensors in their noses. With the young puppies' bodily functions not yet fully operating, mothers lick the anogenital region of their offspring to stimulate the elimination of body wastes. Like wild canid mothers who wish to keep the presence of their defenseless pups and cubs hidden from predators, the mothers of domestic dogs also instinctively consume all of their puppies' waste products.

For the first two weeks of their lives, puppies are virtually comatose. They can't even wag their tails or raise their ears. But this placid exterior belies what is going on within: Their brains, bodies, and central nervous systems are developing at an astounding pace. At around two weeks of age, a pup's eyes and ear canals open. Loud noises will now frighten what was previously an oblivious dog. At first, puppies see only light. Gradually, they begin to detect movement and discern shapes. By four weeks of age, their vision is almost equal to that of an adult dog. Only the ability to see at a distance remains poor. In another week or ten days, this too will have fully developed. Around the same time, domestic and wild pups learn to stand and, shortly after, to walk. Their tails can now rise to the occasion and help provide balance as they move about. Armed with these new and improved sensory and physical capabilities, puppies, both wild and domestic, begin exploring their environment. By now, their ability to regulate body temperature is fully functional, and they can leave their mother and littermates behind for short sightseeing expeditions. By five weeks of age, canids can run.

Mothers continue nursing their young charges until the arrival of the first teeth at anywhere from three to six weeks of age, when weaning begins.

In wild canids, the young learn to lick their mother's face and throat to stimulate regurgitation. Partially digested, regurgitated food is easier on young digestive systems not yet used to breaking down solid foods. Although this behavior is still seen in domestic dogs, human intervention has succeeded in eliminating it to a great extent. By this age, some owners may have introduced semi-solid foods like baby cereal or a gruel made from ground-up puppy food and water or milk replacer; others will have gone straight from milk to commercial puppy food.

With appetites stimulated by their increasing physical activity, which includes playing and roughhousing with their littermates, pups become prodigious eaters, consuming double the amount of solid food that an adult of the same size—but of a smaller breed—would eat. This exercise and extra food intake improves strength and coordination. By three months, pups can walk and run almost as well as adults.

Between twelve and sixteen weeks of age, the pups enter what is known as the juvenile period, or pre-adolescence, lasting until they are roughly six months old. Although their weight will likely have tripled by this point,

Between four and six weeks of age, both domestic and wild pups, like these rambunctious young coyotes, are strong and coordinated enough to engage in some rough-and-tumble play-fighting with their littermates. By ten to twelve weeks of age, fully weaned and strengthened by solid food, wild canid pups start wandering farther and farther from their dens.

At about five or six months of age, these young wolves are still at least six months away from the complete physical maturity of their mother. Fully grown, gray wolves can measure up to three feet tall at the shoulder and weigh as much as 175 pounds. Females of the species generally weigh about 20 percent less than males do.

A DOG'S LIFE

Actual age	Equivalent age SMALL BREEDS	Equivalent age LARGE BREEDS
1	15	12
2	24	19
3	28	26
4	32	33
5	36	40
6	40	47
7	44	54
8	48	61
9	52	68
10	56	75
11	60	82
12	64	89
13	68	96
14	72	103
15	76	110

Despite the widely held belief that dogs age seven years for every actual year, their aging process is a little more complicated than that. As shown in the chart above, the size of a dog plays an important role in its life span.

they will continue growing throughout their adolescence and into early adulthood. Most wild canids and domestic breeds complete their skeletal growth by ten to twelve months of age, while the giant domestic breeds may keep growing until they are eighteen months old.

GROWING UP

Marked by the onset of puberty and its raging hormones, most canids enter the turbulent period of adolescence at about six months of age. Flooded by an onslaught of testosterone, male dogs in particular begin discovering their own sexuality. They may have undergone some emotional and sexual changes early in life, but won't reach true sexual maturity until much later. A release of sex hormones just before birth is said to "masculinize" the male dog's brain; his testes descend into the scrotum before he is a month old. And even though they may display mounting and other sexual behaviors at a very young age, males—like females—only become sexually mature and capable of reproducing after they have reached puberty. Their newfound sexual drive will likely inspire frequent roaming, aggression toward other males, and urine marking (*page 52*) as the males begin seeking out available females for mating.

Male canids are sexually active throughout the year. Prompted by seasonal changes and the production of the hormone estrogen, most domestic females or bitches undergo two periods, called estrus or heat, per year in which they are interested in sex. Sexual maturity comes later to wolves, with females experiencing their first estrus after the age of two, while males mature at around three years. Female basenjis, like wolves and most other wild canids, have only one heat per year.

THE FINAL YEARS

Although some canids don't become emotionally mature until they are almost three years old, most reach adulthood at eighteen to twenty-four months. Among domestic dog breeds, the aging process varies widely, usually depending on the dog's size (see chart, opposite). Some of the giant breeds, like Irish wolfhounds, may lead relatively short lives, their over-worked hearts sometimes expiring as early as six or seven years of age. But apart from specific aging-related illnesses such as diabetes and arthritis, older dogs in general begin to encounter the familiar and gradual deterio-ration in physical abilities wrought by time. Like humans, the middle-aged dog's girth may increase and the limbs may lose muscle tone. As time goes by, hair might turn gray and the senses may begin to dull.

Medical and nutritional improvements over the last two decades are thought to have extended the domestic dog's life by up to 50 percent. Meanwhile, wild canids have not been so lucky. While healthy domestic dogs, especially the smaller breeds, sometimes attain ripe old ages of eigh-teen or twenty, their wild relatives rarely reach half of that. Although wild canids in zoos have been known to live up to fifteen years, hunting pres-sures, outbreaks of disease, and predation by other carnivores have con-spired to keep canid life spans in the wild under ten years, and in some species, even less. Life in the wild, it seems, is not only nasty and brutish, but all too often, short.

A pair of domestic dogs are locked in a post-coital embrace by the swelling of the base of the male's penis. Lasting about thirty minutes after copula-tion, this prolonged coupling technique, or "tie," is thought to improve the prospects of fertilization. Although most male and female domestic dogs achieve sexual maturity around six months of age, some of the giant dog breeds take as long as eighteen months to mature. Most domestic females have two periods of estrus per year, the onset of which is thought to be influenced by geographic latitude and its effects on seasonal changes. Like wolves and dingoes, basenjis have only one estrus. They hail from a geographic area where no significant seasonal changes take place.

DOG
BEHAVIOR

· · · ·

**"A dog can express more
with his tail in minutes
than his owner can express
with his tongue in hours."**

·

ANONYMOUS

PACK
·MENTALITY·

Domestic dogs have inherited some of the sensitivity to the pack mentality and hierarchy that help keep the peace in wolf packs. As with wolf pups, the communicative behaviors signifying dominance and submission are often incorporated into dogs' play behavior. Whether the black dog on its back in this photograph is just having fun or is trying to appease dominant dogs would depend on whether this scenario was preceded by play bows (page 94) or by aggressive threats by the standing dogs.

Opposite: Wolf packs are led by a dominant male and female pair known as "alphas." Pack life is organized along gender lines as well as hierarchical ones: The male alpha dominates and disciplines his male subordinates and will fight any male interlopers, while his mate does the same with her female charges and any enemy wolves of her sex.

Overleaf: Gray wolves

The wolf has bestowed a great deal upon its descendant, the domestic dog, including speed, strength, and keen senses. But perhaps the greatest gift of all passed down from *Canis lupus* to *Canis lupus familiaris* is not a physical legacy but a behavioral one. Pack mentality, or the ability and willingness to live life as part of a group, is undoubtedly the most important factor in the domestic dog's long and successful relationship with another pack animal: humans. Today, an estimated seventy-four million dogs around the world share their lives with people—an enduring testament to a way of life forged by canid ancestors millions of years ago.

SOCIABLE CANIDS

Although there are species such as foxes and the maned wolf in South America that live solitary lives, most wild canids are pack dwellers. What causes these creatures to live their lives communally? Prey size and the need for cooperative hunting are certainly important factors. To take down prey larger than themselves, like moose or wildebeest, species such as wolves and African wild dogs must hunt in groups of at least two and usually many more. But even where prey is small and easy to take down, and hunting can be carried out alone, canids such as coyotes, jackals, dholes, and bush dogs live together. They reap other benefits from pack life, including the ability to defend themselves, their recent kill, and their sometimes large territories from other predators.

Pack life also helps when caring for recent offspring. Wolf and jackal packs employ what is known as the "helper" system, with adults and yearlings sharing in the feeding and protection of the young. In the case of wolves, females who have not given birth may still become "pseudopregnant," experiencing hormonal changes that enable them to lactate and provide food if something should happen to the breeding female.

The size of canid packs depends on several factors, most notably the size and abundance of prey *(see box, page 48)*. Wolves live in packs of anywhere from three to thirty individuals. Most frequently, the groups number between six and ten, but the size of a pack often will change. Wolves will hunt in packs when ecologically necessary. However, because their diet is not restricted to large prey, and because they are omnivores, they can also scavenge, capture small prey by themselves, or eat things other than meat. This adaptability is the reason for the success of the wolf; if the environ-

STRENGTH IN NUMBERS

 Pack size varies sharply from one species to another. Jackals and coyotes feed on smaller animals and are omnivorous, so their packs tend to be small. Others require a greater group effort to bring down larger prey. Like wolves, the Asiatic dhole usually hunts large prey in five- to twelve-member packs. Around mating season, they sometimes form groups, or clans, and hunt in numbers ranging from forty to 100 individuals. However, when only smaller prey is available, smaller packs are more efficient; the fewer mouths there are to feed, the better.

Environmental and human pressures have combined to dramatically reduce the sizes of packs in some species. Only a century ago, African wild dogs may have roamed in groups comprising several hundred members. Today, they are commonly found in groups of six to fifteen. The sharp decline—some scientists have suggested the species may be extinct in twenty years—has been traced to hunting by humans; the increasing scarcity of their prey, such as impala and kudu; and their susceptibility to the deadly diseases of domestic dogs in the area.

A pack of African wild dogs feeds on a recent kill. While the number of members varies in a group of this species, the preponderance of male versus female members in a pack remains constant. For some unknown reason, most newborns of the African wild dog are male.

ment doesn't support a large group, as can happen seasonally, the pack can split up and the individuals can still survive. Typically it is the young wolves that leave the pack and seek out their own territories. These lone wolves may eventually establish their own packs when a male and a female meet, mate, and create offspring. Or they may live out their lives as transients, covering hundreds of miles and avoiding the territory of other wolves, occasionally attempting to infiltrate a new pack.

RANK AND FILE

Wolf packs, like human families, are hierarchically structured. This has obvious benefits. With the rank of each wolf known by all, competition within the group for food and mates is lessened, helping to maintain order and focus individual efforts on a common goal—survival. The result is an established social structure that minimizes in-fighting that might otherwise deplete the overall numbers and weaken the group.

Every pack is directed by a dominant and physically superior male and female pair called alphas, after the first letter in the Greek alphabet. These two control all aspects of pack life, from eating, playing, and sleeping protocol to the administering of discipline. Only the two alphas are allowed to mate, although in some instances, the alpha male will stand aside and let his second-in-command—the beta—mate with the alpha female. The rest of the wolves in the pack have their own, lower-ranking positions on the hierarchical ladder, although these positions can fluctuate. At the very bottom rung, most large packs have a submissive member, called the omega, for the last letter in the Greek alphabet. Mercilessly picked on and dominated by all of the other pack members, this scapegoat must follow far behind the others as the pack travels over its territory. In many cases, omegas leave on their own or are forcefully expelled to the fringes of the pack, although on rare occasions they will move up within the ranks as a new omega takes their place.

Age also plays its role in establishing rank within packs. The alphas are almost always the oldest pack members. In turn, their own siblings and oldest offspring occupy the different positions, from the beta to subordinates to juveniles to the lowly omega. Young wolf cubs (juveniles) are constantly dominated by the adult members of their pack. They learn their place in the

After a kill, the male and female alpha wolves will eat first, and the lowest ranking individual, the omega, will eat last—or not at all. Lower-ranking wolves may compete with each other for food resources, but not with the higher-ups in the pack. Domestic dogs are less sensitive than wolves in situations such as these where dominance might be expressed. While you wouldn't want to try to take food away from a wolf while it is eating, many domestic dogs will allow their food bowls to be removed. Some dogs, however, show a sensitivity similar to their ancestors and become aggressive if someone approaches their food, which can become a serious behavior problem.

Hardy breeds such as Alaskan Malamutes and Siberian huskies are often employed as sled dogs. The ability to coordinate their efforts and to follow the directions of both a canine and human leader allows these sled teams to pull their loads quickly and efficiently. They are the ultimate domestic group.

social scheme of things at an early age. Later on, these juveniles will dominate the next generation before moving on and starting their own packs, or in some cases, becoming the leaders and breeders of their current packs.

Pack hierarchy is maintained and reinforced through a highly evolved system of facial expressions and body language *(page 60)*. Even domestic dogs employ some of these expressions and postures with each other and with their human masters. With wolves at feeding time, during play, or at any time throughout the day, regular displays of dominance and submission are used to reinforce the social order and to keep the more rambunctious wolves in line. Staring eyes, assertive posture, and even growling tell subordinate wolves who's boss. Submissive wolves, in turn, will avoid eye contact, cower, and grovel in the presence of a superior. In less subtle displays, a dominant wolf will assert its superiority over an underling with a head or paw placed on its shoulder or muzzle, or by grasping the other's muzzle in its mouth and forcing its rival to the ground.

The wolf pack's social structure is by no means static. Power struggles for rank advancement are common and often triumphant occurrences, especially in the pack's lower echelons. On occasion, aggressive male subordinates may attempt leadership coups to dislodge the reigning alpha. Repeated displays of insubordination toward the leader often lead to a fight. Intra-pack wolf fights rarely lead to serious injury or death, although they may appear—and sound—ferocious. Once settled, these tests of strength and superiority usually signal a return to normal pack life. Sometimes, however, an unsuccessful challenger will be banished from the pack. And on the rare occasions that an upstart can dominate the alpha, a change in leadership occurs with very little disruption to the pack; the new leader takes over and the dethroned one assumes a subordinate role.

THE SOCIAL LIFE OF A DOG

Unless it remains with its mother or littermates for life or moves to a home with at least one other dog, a domestic puppy's natural canine family life lasts only from birth to the time it leaves to join a human family—usually between seven and twelve weeks of age. But even this short time is sufficient to establish the rules, rituals, and importance of social life in a young dog's developing psyche.

Puppies learn from Day One that their mother is—at least for now—top dog. She directs her litter the way an alpha runs a wolf pack, controlling security and, most important, food. Like the subordinates in a wolf pack, littermates may begin competing for resources with each other—in this case, jockeying for position to find the most productive teat to nurse on. At about three weeks of age, when their physical development permits it, puppies begin to play with their littermates. Through play, they practice how to behave with other dogs. Should play become a little too rough for their mother's liking, she may reprimand one of her charges with a nip. When they are between five and eight weeks old, puppies begin to experiment with dominant and submissive behavior, sometimes trying out a little of each and practicing some of the same body language used by their mother. Signs of a dog's emerging character are usually visible by this age. If the pups are still together at eleven weeks, they usually will have worked out the respective rankings in their group's hierarchy.

By the time it is old enough to be adopted by his new owners, a puppy's social mentality is firmly established, and its allegiances can easily be transferred to a human family. This transition period is crucial. Because people—usually adults—control food and security, dogs will look upon them as the leaders. Now is the time to begin making very clear what is and isn't allowed, through basic training *(page 100)* and by providing structure for the dog. Physical punishment is clearly not the way to deal with a dog's transgressions, and force or harsh methods will not result in a subordinate dog. Instead, early obedience training is the key to deterring any challenges to your authority. Positive, non-confrontational training techniques using food rewards and praise will remind your dog that you are in charge. Continuing these practices consistently—and making sure all members of your family do the same—throughout the dog's life helps ensure a mutually rewarding relationship between you and your pet.

In a typical submissive gesture, a lower-ranking wolf licks the muzzle of a dominant wolf. This is probably a carryover from a juvenile behavior, where the young will lick the face of their mother or other older individuals to stimulate the regurgitation of food.

MARKING
·BEHAVIOR·

When it comes to urine marking, trees are a favorite target of both domestic and wild canids. Although wolves, foxes, coyotes, and other wild dogs mark territory to claim the resources within, domestic dogs have no such practical plan. Although some dogs may, in fact, be marking their territory, most simply seek out messages from other dogs—and leave their own in turn—because they enjoy it. An impatient owner who yanks the dog away when he stops to stiff at messages left by other dogs deprives the dog of a major source of pleasure.

Dog communication obviously lacks the sophistication of human speech, but that doesn't prevent canids from getting their message across in a variety of ways. In addition to vocalizing (*page 57*) and body language (*page 60*), all members of the canid family use scent and sometimes visual markers to make statements to both friend and foe. Where humans build fences to mark off what is theirs, wolves, coyotes, foxes, and other wild canids spray urine, defecate, and, to a lesser extent, scratch the ground to lay claim to physical territories and everything within them— especially prey. Wolves mark the borders of their territories as much as two or three times more often than anywhere inside the boundaries to warn intruders that the space they are about to enter is already occupied. At prominent locations within their territorial boundaries—trees or large rocks, for instance—wild canids leave similar clues, mostly to alert themselves that they're back on home turf as they return from their search for new prey. Like the system of rank that seeks to eliminate fighting between wolf pack members, marking territory helps wild canids avoid the violent and potentially deadly conflict that would otherwise occur between territory owner and innocent interloper. Trespassers detecting the scents and signs of ownership usually know enough to keep out.

"ROVER WAS HERE"
Domestic dogs exhibit marking behavior, although they have no need for the prey that marking territory claims. Nobody can say for sure exactly what practical reason, if any, makes them do so, but many dogs, especially males, leave information-filled calling cards wherever they pass. Much like teenagers with felt-tipped markers scribbling notes on the walls of the high school lavatory, a dog may simply feel compelled to tell others, "I was here."

Because the dog's sense of smell is so refined, scent marking is a particularly effective form of communication. Bodily wastes—in fact, all secretions, including ear wax—produce chemical scents called pheromones that say a lot about their depositor. One sniff gives the receptor a wealth of information, including the age, sex, rank, and reproductive status of the scent's originator. Like a personal ad for canines, a dog's scent message might read something like, "Dominant, single male seeks receptive female for reproduction."

SOLID INFORMATION

Dogs' feces perform double duty as both scent and visual markers. Wild and domestic canids alike often deposit their droppings in prominent locations. The solid waste is a powerful olfactory signpost, packed with useful sexual and social information about the depositor. It is covered by mucus from the large intestine and by secretions from the anal sacs, small reservoirs on either side of the anus and in the nearby perineal glands. To draw additional attention to these already conspicuous markers, male and female canids may scratch at the ground around their fecal deposits, as if drawing arrows for others to follow. These marks may also contain scents from glands between the dog's toes.

URINE MARKING

As anyone who has ever walked a dog (especially a male dog) can tell you, urine marking is an undeniable part of canid communication. The produc-

Wolves not only mark trees with their urine. In some cases, they also leave their scent on a recent kill. A message to other wolves to stay clear of their food supply? A signpost to help them return later when they are hungry? Scientists aren't sure of the answer.

Depending on the size of a pack and the density of prey in a given area, a wolf pack's territory can range from as little as twenty or thirty square miles to up to three or four thousand. Many domestic dogs live in areas dense with potential rivals, and although they may not be claiming territory outright, they are still driven to scent mark almost constantly, especially when encountering a spot marked by another dog.

tion of testosterone between the ages of about four and ten months activates territorial urine marking behavior in adolescent male dogs. Unlike normal urinating for the purpose of relieving the bladder, dogs mark objects with only small spurts of urine and usually keep a little in their reservoir, although particularly fervent markers may run dry after a long and busy walk.

In the wild, dominant wolves sometimes urine mark every three minutes or every three hundred yards (274 meters) or so as they travel along established trails. Male domestic dogs on familiar walking routes have been observed marking up to eighty separate sites an hour. Overall, male domestics will urine mark more often than females. Because they live in close proximity, urban and suburban dogs mark significantly more than wild canids or even rural domestic dogs do. If they're not covering another dog's mark, then they're re-marking one of their own previous deposits to reinforce their presence. Both wolves and domestic dogs ritually re-mark their favorite spots every day or two.

When targeting specific objects, most adult male canids cock one of their hind legs while urine marking to ensure that the deposit is high enough for other canines to notice. In what is known as double marking, a male wolf will mark over a female's urine mark as part of courtship or "pair bonding".

Females and puppies usually squat to urinate, although some female canids employ a variation of raised leg urinating in an attempt to aim the flow. While partially squatting, they lift a rear paw and swivel their rear end in the direction of their target. Lone wolves and lone coyotes leave no scent markers, seeming to prefer to retain their anonymity and not alert others to their presence.

Neutering is the most effective means of nipping the marking behavior of domestic dogs in the bud, so to speak. Once neutered, dogs will urine mark outdoors considerably less often than they did previously and usually will stop marking indoors, if they ever did so, altogether. A male dog who is neutered before learning to mark may never exhibit this behavior at all.

LOVE POTION NO. 9

When females urine mark, it's usually to broadcast sexual receptiveness. Female wolves increase their rate of urine marking significantly as breeding season nears, and domestic females are no different. The pheromones found in the urine of in-heat females contains information about their sexual readiness. These scents can be picked up by the vomeronasal organs *(page 32)* of sexually experienced males, enticing them to mate. Estrus urine will most certainly draw the attention of sexually immature males, too, but their lack of experience prevents them from making a connection between the smell and the female's readiness to mate.

Besides the territories they may be marking outside, most dogs have a smaller personal space indoors to which they retreat for security. In the case of this smooth-coated collie pup, like many other domestic dogs, the favored space is often a basket bed. When comfortable in their home environment, well-trained dogs will usually confine their marking behavior to the outside. However, the presence of another dog or the hormonal changes brought on by adolescence may cause a dog to assert itself by urine marking indoors.

DEFENDING TERRITORY

After companionship, security is the main reason people decide to get a dog.

Defending territory or property is the domestic dog's inheritance from its communal-living wolf ancestors— even if there is a trade-off of forest acres and sprawling, open land for a modest family home or car in a parking lot. Depending on the breed and the dog, this protective tendency may range from very intense to completely absent.

When domestic dogs perceive danger and bark out a warning, they are echoing a wolf behavior. Members of a wolf pack sound the alarm together but subordinate wolves, like domestic dogs, typically step back to let the pack leaders handle the active defense. Of course, many dogs bred for their aggressive tendencies, such as the Rottweiler, will actively defend their turf.

A visitor—or the mail carrier— might think twice before encroaching on this dog's turf.

THE VOCAL
·DOG·

The bark of a dog is one of the most inescapable reminders of canid self-expression. It is also, according to many experts, a legacy of what was once only a minor behavioral characteristic, bred into full-blown prominence by careful selection. When humans began breeding the first dogs from the tamest of wolves, they recognized the fact that barking served as a useful alarm. So, as one theory goes, the most vocal wolves were mated to produce the world's first guard dogs. Today, like a cat's meow, barking serves a varied role in communication, giving voice to a range of emotions and information.

CALL OF THE WILD

In the wild, vocalizing is considered to be a canid's least important communication tool, used mainly to send long-distance messages. Nevertheless, it is a surprisingly sophisticated and flexible system that serves many practical purposes, especially among more social canids.

The most common means of vocalizing in wild canids is howling. Wolves in particular howl for many reasons: to call the members of a pack together for a hunt, to claim territory, to warn of a territorial invasion, or seemingly just for the fun of it. Like a human family's sing-along, communal celebratory howling strengthens the bonds between pack members and gives them a feeling of pleasure and greater security. Both wolves and coyotes, whose higher-pitched yip serves the same purposes as the haunting howl of *Canis lupus*, call more often in the early evening, their most active part of the day.

Besides howling, the domestic dog's wild ancestors employ a number of other vocalizations. Like domestic puppies, young wolves whine, whimper, and yip to attract their mother's attention. As adults, they employ menacing growls to confront outsiders or to reinforce their dominance over underlings within their pack. Other species of canid, like the ferocious dhole, or Asiatic wild dog as it is also known, rely on more unusual utterances. To coordinate their efforts while tracking large prey, these animals communicate with each other through a piercing call that earned them the nickname "the whistling hunter."

Even some of the solitary canids put their vocal chords to regular use. In the winter months preceding mating season, when the conditions for scent marking are poor, some territorial male foxes may use aggressive and raspy

The owner of this akbash no doubt has learned over time to interpret his dog's different barks; most dog owners can, easily differentiating between, "I want to go out" and, "Somebody's out there," as well as other barks conveying happiness, annoyance, or even fear.

Opposite: Coupled with their acute sense of hearing, howling helps wolves communicate over exceptionally long distances. Despite what is commonly believed, wolves don't howl at the full moon any more than they do at the midday sun.

The idea that dogs howl to music because it hurts their ears or otherwise displeases them is a myth. More likely, they howl along for fun, in much the same way that wolves and some domestic breeds do communally for sheer pleasure. If the sound of singing or music were truly disturbing to a dog, it would simply leave the room.

bark-like calls to keep potential rivals at bay. At close quarters, they emit a shrill scream for much the same purpose. To alert her pups to a potential danger, a female fox, or vixen, may employ a soft warning bark.

Despite what many people think, wolves do indeed bark, especially when strangers—lupine or otherwise—approach the pack. Softer and significantly lower in volume than the domestic dog's bark, the wolf's unique vocal alarm probably galvanized early humans to begin breeding them for this, among other behavioral traits.

NOISY BARKERS

Domestic dogs are considerably more vocal than wild canids. Because their natural social life and territorial needs have been seriously modified by domestication, they howl considerably less than wolves, but make up for it by barking a lot more. Semi-domestic or primitive dogs like dingoes and New Guinea singing dogs—named for their curious, almost plaintive sing-song howl—fall somewhere between domestic and wild canids. The primitive basenji of Africa is often referred to as the silent, or barkless, dog. In reality, it has a high-pitched, yip-like bark and a howl that sounds more like a yodel.

Domestic dogs have a vocal repertoire consisting of several different varieties of sounds. Some use infantile cries, whines, and whimpers even as adults to beg or plead attention, affection, or food from humans. Perhaps by providing for dogs throughout their lives, we have extended into adulthood their puppyhood, or neonatal period, with its associated vocalizations. Adult domestic dogs almost never utilize these sounds of infancy on one another. If it is not rewarded, this infantile vocalizing can be controlled and perhaps even diminished.

Sharp barks and growls constitute warning or alarm sounds. Barking in itself can mean a great many things (page 95), from joy to anger to curiosity, so the possibility of misinterpretation is high. Often, barking indicates internal conflict as dogs vacillate between duelling emotions such as, "I like you," and, "I'm a little afraid of you." Plaintive barks and howls make up attention-seeking sounds. Dogs employ these vocalizations to call humans or other dogs and when they are suffering from separation anxiety, although they are often mistaken as cries of boredom.

Certain domestic breeds howl more than others. Like wolves, the so-called northern, wolf-like breeds such as Siberian huskies or Alaskan Malamutes howl communally to strengthen the bonds between pack members. Like hound dogs, whose baying call helps them coordinate a hunt and who howl in celebration afterward, sled dogs also may celebrate after a race.

The most easily interpreted sound dogs makes is a yip, yelp, or scream of pain. Although it can sometimes signify emotional distress, this type of vocalization is most often a sign of physical pain. Moans of pleasure are learned sounds that we encourage by offering belly rubs and other forms of affection. Like infantile sounds, they are likely reinforced by humans, and are never heard in the wild.

Individual dogs possess a unique voice, a vocal fingerprint of sorts that is indiscernible by human ears. For the most part, of course, the larger breeds have deep, resonating barks, while the smaller breeds make themselves heard in higher tones. There are, however exceptions to this rule. The relatively small cavalier King Charles spaniel, who tips the scales somewhere between ten and eighteen pounds, has a surprisingly loud and deep, booming bark.

Small dogs are by far the most talkative, yipping away with shrill barks. Or, like the Pomeranian, they may have originally been bred as watchdogs. It is possible that owners of small dogs unintentionally reinforce this often grating habit by continually picking up these tiny yappers to calm or comfort them, encouraging the undesirable behavior in the process.

Excessive barking by any dog can be a real concern, and may lead to serious disputes between human neighbors. Training will often keep the worst barkers in check (page 117). The alternative that some owners take is surgical de-barking, a technique that was once used during World War I to silence dogs that carried messages back and forth in the trenches. For dog owners of today, however, this measure makes little sense and robs dogs of one of their principal outlets of self-expression.

Small dogs, like this aggressive-looking toy fox terrier, are usually the most vocal of domestic canines, seeming to make up with volume and persistence what they lack in size.

The dhole emits a howling whistle to communicate with its brethren while they circle prey in the undergrowth, enabling the group to coordinate the attack.

BODY
·LANGUAGE·

With wolves, as with domestic dogs, body language can easily be misinterpreted. In the photo above, it might appear that the wolf on the right is asserting its dominance over the apparently submissive wolf on the left. In fact, the omega wolf of the pack is grabbing the tail of the alpha wolf—most likely a play behavior, since an omega would rarely challenge the dominance of an alpha in such a blatant fashion.

Two wolves confront each other. For a second, their eyes meet, then one averts its face and lowers its ears, tail, and body. The other stands tall, ears and tail up, staring directly at its adversary. The human viewer knows something has happened, but what? Mostly in silence, the two wolves have transmitted a clear message to each other. The first has indicated that it is submissive; the other has asserted its dominance. This is a scenario played out continually in the world of wild canids, and to a certain extent with our pet dogs, who have retained some of these communicative behaviors. However, the body language of domestic dogs carries a different meaning than it does in wolves.

Wild canids do, of course, vocalize *(page 58)* and leave distinctive scent messages *(page 52)*, but body stance and posture also play an important role. They read other canids by the set of their tails and ears, and the appearance of their eyes and mouth. The dominant individual sends messages of confidence: body four-square on the ground, leaning forward, ready to act, daring the rival to maintain its gaze. The submissive one may acknowledge the other's dominance if sufficiently threatened by rolling over and baring its underbelly as if saying, "I am no threat." Most dominant canids will accept this obeisance as their due, perhaps giving a sniff to the submissive one's genital area, but otherwise not threatening it.

In the wild, an alpha wolf can often maintain its preeminent position in the pack by a fixed stare alone. But it can also call on other techniques, puffing itself up in size physically and psychologically by raising its tail and ears. When two dominant wolves meet, they will stare and posture, showing teeth and raising hackles, until one backs down or a fight ensues. Almost invariably, the body language of submission allows such confrontations to end without physical damage to either party and permits a wolf defeated in battle to "cry uncle" without further harm. The order of the pack is affirmed, and each wolf knows where it stands in the hierarchy. Other creatures like foxes and dholes may rely on subtle changes in their mouths, ears, and tails to convey a variety of meanings to members of their kind.

Some domestic dogs are more attuned to this system of canid communication than others, while some may be almost oblivious to it. One study has shown that the messages of dominance and submission that dogs display to one another through body language are greatly diluted from those seen in adult wolves, and more closely resemble the behavior of juvenile wolves. Although some dogs may go through the motions of the dominant-submissive ritual, few of them actually engage in more serious contests of power. These behaviors are more likely to show up during play, where they have a different meaning. Domestic dogs usually have no need for true aggressive or submissive behavior, since their survival doesn't depend on it. The small percentage of domestic dogs who are sensitive to the body language used by wolves and other canids to diffuse power conflicts in the wild can become aggressive toward other dogs or people if their signals are ignored or misunderstood.

A dog's gestures can convey other messages, such as interest or a desire to play, to canine comrades. But people can easily misread these clues, and a dog may have more than one inclination at any one time. For instance, many dogs can swing from docility to anger with no change in their body language. Humans need to be careful when decoding a dog's physical cues *(page 93)*.

In a display of dominance, the dog on the right stands over the other dog, with raised ears and tail, staring intently. The dog on the left, lowered into a submissive position, averts its eyes and holds its ears and tail down. While similar to the posturing of wild canids, this body language usually occurs in play with domestic dogs, and in most cases ends up with the two frolicking together.

COMMUNICATION BREAKDOWN

The combination of selective breeding and cosmetic surgery molds dogs to suit human tastes, but such modifications can have an unexpected consequence: miscommunications among canines.

When dogs are bred for heavy, long coats, for example, other dogs have difficulty seeing their eyes, ears, mouth, and raised hackles and the messages they normally convey. Surgically altering a dog's ears to remain erect and forward means that it will look perpetually alert and dominant, regardless of its true personality. And docking a dog's tail eliminates one way of conveying its feelings to fellow canines.

While part of its scruffy charm, the long hair of a bearded collie can mask some essential communication cues.

LEARNING AND
·INTELLIGENCE·

The intelligence of the Border collie makes it an ideal breed for challenging tasks. If you have one as a family pet, make sure he gets constant stimulation and exercise to prevent the difficult and destructive behavior that can result from his boredom.

L ike humans, dogs spend a lifetime gathering knowledge. And while the learning rate may slow down as a dog ages, you *can* teach an old dog new tricks. It's just that dogs retain information—emotional and experiential—more quickly at a younger age when their minds are more impressionable and less cluttered. Eventually, a dog's cognitive abilities diminish and the animal's brain actually shrinks in size. But even at an elderly age, canines are capable of adapting to their environment and adding to their repertoire of skills.

THE INTELLIGENCE OF DOGS

The subject of dog intelligence elicits various tales from owners, often prefaced by, "My dog is so smart that. . ." How much any of these examples are related to intelligence is a moot point. To a great extent it depends on the definition of intelligence. With humans, intelligence is often associated with problem solving. But of course there are many different kinds of problems. A herding dog that can corral dozens of sheep at a time, a wolf that learns to raise a trapdoor by pulling a rope, a fox that foils a chicken coop owner's best efforts to keep predators at bay, a guide dog that has to find a safe route to lead its owner through a new construction site—all offer compelling testimony to relatively agile minds at work.

This doesn't mean that canids think like humans. Scientists who study how other animals think doubt that canids reflect on their behavior. For example, although people often say, when a dog misbehaves, that he "knows better," this isn't the case. Dogs are not likely to think back—even a few moments—about past behaviors, nor are they likely to think ahead about the consequences of current behavior. Their world is rooted in the here and now. Believing otherwise is a common cause of owners' misinterpreting dog behavior, although it is very easy to make such a mistake. Dogs often look "guilty" when the owner approaches or discovers evidence of some undesirable act, leading the owner to assume "he knows what he did." But this would require that the dog reflect back on what he might have done that now makes his owner angry. Most often, this guilty demeanor means the dog knows that in the presence of certain signs, such as his owner's body language and voice, and under certain conditions such as when there is a pile of shredded newspaper or a puddle on the floor, he will be punished. But the fact remains that he is unlikely to know *why* he is being punished.

As humans, we are constantly making inferences about the behavior of other humans, such as what motivates their actions toward us, or what their next move might be. We begin from the assumption that the mind of others works like our own. Although this may work effectively in our interactions with other humans, these assumptions lead us astray when we are trying to understand a dog's behavior. Often, owners assume when their dog does something inappropriate, like chewing the sofa or eliminating on the rug in the living room, that the dog did this because he was angry or jealous or spiteful. But for an act to be spiteful, a dog would have to think about how his actions would affect humans at a later date. Scientists who study cognition don't believe dogs can do this.

For a more accurate understanding of any domestic dog's behavior, we need to strip away those rather automatic inferences we are in the habit of making for human behavior and reframe the scene from the canine's point of view. Much of any dog's response is instinctual, such as wagging the tail in welcome. Still, dogs show the ability to discriminate: They don't wag their tails at everybody—only those whom they identify as friendly. A dog's logic is a simpler logic than humans'. Dogs do not think abstractly or make intuitive associations at the level at which human beings make them.

Working in consort with a shepherd, Welsh corgis like the one shown above can keep their charges moving in the right direction with a combination of eye contact, movement, and the occasional nip. They and other breeds with the ability to learn and perform such demanding tasks are often credited with being among the most intelligent of dogs.

As with people, domestic dogs differ in their intelligence from individual to individual. Of course, measuring dog intelligence presents some problems because, by definition, it is not the same as our own. If we measure intelligence by a dog's value to people, then the smartest dog will be the most trainable. Of course, we can choose another perspective and define dog intelligence in terms of sheer ability to survive under a variety of conditions, perhaps the ultimate test of usable brain power. By this argument, the smartest dog may actually be the independent and intractable stray dog, surviving in the world without human aid.

Of all the canids, the wolf is considered by many scientists to be among the most intelligent, better at problem solving than a domestic dog and more skilled at learning by observation. In one experiment, a human showed dogs and wolves how to open a door by turning a handle. The dogs never learned how, but the wolves quickly mastered the task—not surprising for creatures that must survive by their wits alone.

HOW DOGS LEARN

As all dog owners have discovered, sometimes much to their chagrin, their pets learn constantly—even when nobody is intentionally teaching them. In fact, sometimes we spend a lot of time trying to modify responses that our dogs have learned and we wish they hadn't. Just ask the person who has fed the family pet scraps at the dinner table and then later tries to stop the behavior. Also, different dogs may learn the same behavior in different ways. If two dogs are housebroken in a home with only wood floors and later, after a carpet is installed, one starts eliminating on it, we could suspect that he had learned simply not to go on wooden surfaces, while the other dog had learned not to go inside the house.

Much of what dogs learn is related to the context—where they learn and with whom. A dog may learn that if he starts to explore the garbage when you are standing nearby, he will get scolded. So he learns not to do it then. However, he may also learn it is okay to spread the garbage around when you aren't there because nothing bad happens when he does, and will not understand why he is punished when you come home. Commonly, when family members train the family dog, some will be more effective than others in getting the dog to obey. This occurs because the learning experience has been different with each family member.

In dogs, cats, humans, and other animals, behavior is influenced by consequences. When a reward follows a particular behavior, that behavior is more likely to occur again. If instead something unpleasant occurs, the behavior is less likely to be repeated. Fortunately for us, dogs have inherited from their wolf ancestors a strong motivation for social acceptance and approval. We can take advantage of a dog's desire to please by using attention and praise to reward good behavior. Cats, with a less complex social

Obeying the "Sit" command, this Samoyed receives a treat as soon as his rear end is on the ground. A dog will learn best if he is rewarded immediately for his actions. An important ingredient in the learning process is attention: If the dog isn't paying attention to you, the lesson will not be successful. A tired or anxious dog will have more difficulty concentrating. Distractions also can interfere with learning. A particularly enticing reward can help bring back the focus to the task at hand.

system, can live comfortably alone or as a part of a group, care less than dogs do about our approval, and therefore are much harder to train. When training a dog, if we initially praise a correct response to a command and immediately follow with a treat—a food reward or a few moments of play—the praise will become even more effective as a reward in and of itself. Over time, treats can be reduced and provided intermittently following praise to teach and maintain these behaviors.

Finding what rewards work best for any particular dog is a process of trial and error. A reward should capture and hold the dog's attention and be attractive enough that the dog is willing to work to get it. To be effective, a reward must be provided immediately following the correct behavior, within a second or two if possible. Later than that and you will be rewarding a different behavior. For example, often in the process of housebreaking, an owner will open the door and let the dog outside to eliminate in the yard. The dog runs out, does his business, then runs back in and receives praise and a treat at the door. While the owner believes he has rewarded the dog for eliminating outside, the dog is more likely to associate the reward with coming back into the house.

Sometimes an intended punishment is experienced as a reward. A dog desperate for attention may find it rewarding to be being yelled at or chased. If a dog likes to jump and nip at hands and arms, and the owner gives a firm "No" but at the same time is pulling to get free or pushing the dog away, the dog may find the latter behaviors an enjoyable kind of play. This enjoyment is likely to far outweigh the verbal correction.

Punishment is not a particularly effective method for training a dog. Usually a strong, clear "No" is the most severe correction needed to indicate a behavior that is inappropriate. Harsh punishment is both unkind and an inefficient way to get a dog to learn what you want. A correction that causes pain or fear may cause the dog to inhibit a particular behavior at that moment, but the intense emotions caused by such actions can lead to unfortunate and unintended consequences. Fear or pain may become associated with the owner or with some other aspects of the environment where the correction occurred, leading to serious problems with aggression and/ or anxiety. This kind of learning involves involuntary responses and can be very difficult to undo. If you're unsure of how to achieve the desired results without using harsh methods, seek the help of a good dog trainer.

Short training sessions of several minutes alternating with breaks for the dog to sniff and explore or engage in play are more effective than one long session. Training exercises can and should be fun for both human and dog. This process strengthens the bond between the dog and the trainer—whether a professional or simply the dog's human family member—improving communication so that each learns what to expect from the other. All members of a household who are able can benefit by participating in this process.

Beyond basic training, some dogs can learn to perform more elaborate tricks, like this Pomeranian standing on his hind legs to beg. This level of trainability rises to an even higher plane with some performing dogs on television and in movies, who can learn to carry out almost humanlike tasks on cue from a trainer. These dogs are often chosen from the herding group, which includes breeds such as collies and shepherds—among the most trainable of domestic dogs.

CHAPTER

· 3 ·

CHOOSING

A

DOG

· · ·

"If you can't decide between a shepherd,
a setter, or a poodle,
get them all...adopt a mutt."

·

AMERICAN SOCIETY FOR
THE PREVENTION OF CRUELTY TO
ANIMALS

BEING

·RESPONSIBLE·

Neither sleet nor snow nor rain should keep your dog from his daily exercise. So prepare yourself. You're looking at a minimum of two walks a day, every day, for as long as you're blessed with your furry friend. Depending on the climate and the breed, you may need to outfit your dog with a sweater or waterproof boots.

Overleaf: Mixed-breed golden retriever

What are you doing for the next ten or fifteen years? This may seem like a silly question, but it's the first thing you should ask yourself when you're considering bringing a puppy home. Even if you're thinking of adopting an older dog, prepare to dedicate many years to this newest member of the family. You'll have to fit a dog into every aspect of your life, including the possibility of relocating to another neighborhood, city, or country, moving in with another person, going on vacations, having children, and perhaps dealing with the stresses of a family member's illness. This is not a decision to make on a whim.

While dogs need more daily care and attention than, say, cats, your dream of having a dog shouldn't be squelched by the fact that you have an active lifestyle—as long as you pick the appropriate breed. Although it's essential that you spend some time with your dog every day, some breeds require less chore-type care and attention than others. Once you've considered these factors, and you're confident that yours won't be one of the innumerable dogs left at animal shelters each year, you're one step closer to living with one of the most loyal, loving friends you'll ever have.

ARE YOU PREPARED?

If you're not sure that you have enough free time to dedicate to a dog—including two walks a day, play time, and clean-ups—you might consider caring for a vacationing friend's pet. Or try fostering a dog from your local humane society for a week or two to get a better feel for what dog ownership will entail. Still not sure? Maybe you need to mull it over more. Remember, it's easier to wait until your life can naturally accommodate a canine housemate than it is to try to make radical readjustments after adopting a dog.

Keep in mind that dogs require yearly veterinary check-ups and vaccinations and, in some areas, preventive medication—against heartworm, for instance. Also, a puppy will need a whole set of initial vaccinations. If you've taken in a stray older than four months, he'll need an all-in-one shot *(page 159)* plus the rabies vaccine. Should your dog get sick—a normal event in the course of any life—you'll need to provide for proper veterinary treatment and care. Don't forget dog food, toys, crates, collars, leashes, licenses, and other essentials. The point is, all these things can cost upward of five hundred dollars per year, depending on the breed. A puppy's first-year expenses can run to roughly nine hundred dollars. This includes not only food and toys, but also neutering, vaccination, and other items, but not the cost of acquiring the puppy in the first place. Factor these figures into your annual budget if you're going to take the responsibility of pet ownership seriously.

While all pet dogs need lots of affection, the active attention your dog requires will depend as much on his breed as on his personality. Lapdogs, such as pugs and Chihuahuas, need only about forty-five minutes of activity per day, while large working and sporting breeds such as huskies and setters need at least two hours of exercise per day. While all breeds exhibit particular exercise and behavioral characteristics *(page 176)*, you can never be completely certain what your dog will do next. Make sure he is well trained *(page 100)*, secure, and well cared-for. Other than that, play with him often and give him plenty of chew-toys.

Consider your living situation before you bring any dog home. Single apartment dwellers should think twice before acquiring an energetic dalmatian pup, for instance. Never get a dog for your kids unless you want one yourself. Children can help, but parents should be prepared to take on all dog-care duties since youngsters don't always keep the promises they make when they first set eyes on a litter of cuddly pups.

Be mindful of your responsibility to the environment. Simply bagging and disposing of dog waste is no longer enough. Check out one of the environmentally-friendly composting units made for this problem, or use a commercially available mix of water and enzyme powder, which will break down your dog's stools.

Remember, too, that allowing your dog to interact with other animals may be fun for him at first, but can pose a danger. Keep your dog on a leash when stray or strange animals are around, and when he will be in contact with other species. No dog should be let loose on a wildlife preserve. Both wild animals and your dog might get hurt.

CHILDREN'S DOG-CARE RESPONSIBILITIES

AGE	DUTIES
Up to 5	Help to choose and name the dog. Help adult with brushing, and choosing dog toys. Play with the dog; an adult should be present to teach and ensure gentle play.
5-8	Help with grooming; play with dog; help put pre-measured amount of food into bowls, and provide water. Clean up after the dog—all under adult supervision.
9-12	Provide food and water; groom; walk and clean up after dog. Also, should be able to recognize physical problems—such as limping—that may require veterinary attention.
Teen years	Take over feeding, grooming, and walking. Run errands to restock dog food supply, toys, etc. Accompany adult to vet and help to administer any care and medication required.

LIVING WITH ALLERGIES

 Dogs are constantly sniffing and licking to identify things, shedding fur, and leaving scent from their paw pads. If you suffer from allergies, these canine traits can produce uncomfortable, sometimes even fatal, reactions.

One type of allergen is found in dogs' saliva. Another type is made in their oil-producing glands, secreted onto the skin, fur, and paw pads, and then shed with surface skin cells as dander.

While it's true that some people's allergies are too severe to allow them to live with a dog, mild allergy sufferers have a number of options. Antihistamines help, of course, and dog grooming and bathing can be left to other family members. Here are some additional tips:

◆ Buy an air purifier.

◆ Train your dog not to jump up on you.

◆ After touching or playing with a dog, wash your hands and all chew toys.

◆ Bathe the dog often, or swab him with commercially available allergy-proofing products using a cloth.

◆ Try to keep dogs out of areas that you frequent, such as your bedroom. If that's not possible, wash and change bed linens, vacuum, dust, and mop often wearing a dust mask if necessary. Invest in a vacuum with a HEPA (high efficiency particulate air) filter.

◆ If you can't keep your dog off rugs, carpets, and upholstered furniture, consider removing the carpets and rugs and covering furniture with blankets or sheets that are cleaned often.

◆ Wash walls, counters, and other surfaces that your dog may have sniffed or licked.

◆ Wash the dog's sleeping blanket or bed cushions regularly.

THE SELECTION
· PROCESS ·

Big or small, thick or thin, long- or short-haired, and available in a myriad of colors and shades, there is a dog to suit just about anyone. Before you bring one home, keep in mind the size of your family, the living space you have to offer, and the amount of time you have to spend with your dog or puppy.

You've prepared yourself for all the responsibility, cost, and time that living with a dog will entail. The next logical step is deciding on the breed. Perhaps you're looking for a dog who will get along well with your children or other pets. Or maybe you would like a companion on long runs through the countryside. Your first consideration is both the most obvious and the most important: Choose a dog whose space and activity needs match yours. Don't get a Great Dane, for instance, if you live in a small apartment. Trainability should also be a top priority, especially if your dog will be expected to interact with family, friends, and strangers. If you have a problem with messy dog hair, consider a breed that sheds minimal fur, perhaps a bichon frise. Golden retrievers are great with kids, but need plen-

While you know more about what you're getting in terms of look, color, behavior, and intelligence with a purebred dog such as the Rottweiler at left, your dog may also be susceptible to certain illnesses and breed-specific problems such as hip dysplasia. A mixed-breed dog, like the one above, is usually friendly, healthy, and intelligent, but you may never be quite certain of what he will become as he grows and ages. His unknown heritage can be a blessing or a Pandora's box.

ty of exercise. Chihuahuas, on the other hand, don't need much exercise but are generally too fragile for young children. Whatever your need or desire, there's a breed to suit you. Dog ownership is a long and serious commitment, so don't make a hasty decision.

PUREBRED VS MIXED BREED

Although any dog is in part an unknown quantity, purebred dogs are bred to preserve and pronounce physical and behavioral traits. Therefore, almost everything about them can be predetermined—from their potential size and temperament to their life expectancy. While this is usually a good thing—for instance, you're always guaranteed that thick, luxurious coat with a Shih Tzu, and you can be certain that your Doberman will be a great running companion—it can also accentuate certain undesirable traits and increase a dog's tendency towards hereditary disease. If you respect the natural behavioral tendencies of your purebred, and make certain that the dog you choose doesn't come from a closely inbred litter, you should avoid most problems. A vet, animal shelter staff member, kennel club member, respected dog breeder, or good book can help you identify some of the more common problems with different breeds.

Puppies such as this golden retriever have energy to burn and will spend hours playing. A good way to bond with your dog during his peak period of socialization—from three weeks to three months—is to share in the fun.

By definition, the mixed-breed dogs typically found at pounds or animal shelters are of uncertain or unknown parentage, but they usually tend to be intelligent, loving family pets. Of course, there's no real way of knowing how a mixed-breed puppy will shape up in terms of either appearance or temperament when he first comes home. The fate of crossbreed dogs—the offspring of two different purebred dogs—is usually easier to predict, but even this is no guarantee. Mixed-breeds are usually cheaper to buy than purebreds, and are more resistant to many of the health problems to which purebred dogs are prone. The more moderately sized mutt can usually sidestep the hip dysplasia that plagues many larger breeds such as Rottweilers. When selecting a mixed-breed dog, your best bet is to ask the shelter staff or litter owner if they have noted anything unusual about the behavior of your first choice. If so, move on to the next candidate.

Don't choose a purebred dog just to follow a trend. Many mixed-breed dogs are in greater need of homes, since purebred dogs have a higher perceived status in today's society. No matter what your choice, make sure that the seller is reputable *(page 77)*.

PUPPY VS ADULT

Few people can resist the impish charm of a puppy. But don't discount adult dogs in your search for the perfect pet. While puppies are playful and cute, an adult dog presents fewer unknowns in terms of potential health and behavioral problems. Choosing a full-grown dog also means you bypass the

Bringing home an adult dog means you know exactly what you're getting in terms of appearance. Spend a little time with him beforehand and you'll have a good idea of his temperament, too. You may find a fully trained dog, saving hours of training time. Just make sure that you're not taking in a dog whose previous owners gave up because of health, control, or training problems.

considerable initial costs incurred with a puppy—among them, spaying or neutering, initial vaccines, and training classes. In addition, a family giving up their full-grown pet might supply the new owner with essentials such as a cage or crate, leashes, dishes, toys, and blankets, which can add up to great savings. Of course, a puppy's personality may be easier to shape, and behavioral difficulties are easier to eliminate than with adult dogs.

If you choose a puppy, look for one between the ages of seven to eight weeks old; any younger and he will likely have behavioral and social problems later. Once you bring the puppy home, however, you and your family take on the responsibilities of socialization previously held by his mother and littermates, so start bonding right away. This crucial period of socialization begins when a puppy is three weeks old and lasts until he is three months old, though he still needs attention afterward, of course.

First-timers should choose a puppy who is outgoing but not too dominant or aggressive, and may want to avoid the runt of the litter. Runts may require more veterinary attention and training than do the other pups. Give your puppy a quick health check before you bring him home. Look for eyes that are clear and bright, pink gums and white teeth, and a clean and shiny coat. Take note of a puppy with dandruff, ear discharge, or running nose; while these may not be serious health problems in the long run, they will cost you some time and money to correct. Finally, check that he runs and walks well.

An adult dog comes with a known personality and medical history, provided he's not a stray *(page 81)*, or a shelter dog with questionable heritage. If you don't know much about the dog's past, do a few tests to determine whether he is right for you. Take him for a walk through a crowd and watch his reaction to adults, children, and other dogs. Also try giving him a toy to see how easily he lets go. An aggressive dog might snarl, and a meek dog might cower at your approach.

All dogs and puppies have their own built-in requirements and rewards. To ensure that you'll have enough room, time, and energy for your dog, you'll either need to do some creative guessing, in the case of a puppy, or rely on the advice of the previous owner of an adult dog.

BIG VS SMALL

If you have a large house, ample yard, and plenty of hours to spare, a larger breed might fit well into your lifestyle. While smaller dogs generally cost less—food and medications administered by body weight are less expensive, for example—they don't always require less exercise. The huge Great Dane, for instance, needs relatively low-key exercise, while the high-energy miniature pinscher needs a rigorous workout. Nor is size a valid indication of a dog's temperament. The biggest of all breeds, the Irish wolfhound, is a quiet and gentle giant, while small terriers can be surprisingly aggressive and vocal.

Dogs come in all sizes, from the majestic Great Dane, to the mid-sized German shepherd dog and diminutive Chihuahua shown above. Make sure that you choose a dog that fits into your life, whether you want a hardy, active playmate or a delicate lapdog.

The thick, woolly coat of the komondor, originally bred to protect livestock, both insulates it from the elements and shields it from the teeth and claws of predators. With both a soft, curly undercoat and a longer, coarser outercoat, these dogs will become matted if the coat is not properly corded. This is a technique similar to braiding, and is done in lieu of brushing. While the komondor is an extremely high-maintenance breed, even breeds that appear to be low maintenance will require proper grooming to keep them looking—and feeling—their best. If you're not prepared for the time and expense that a breed's coat requires, opt for a different kind of dog.

MALE OR FEMALE?

A wolf or wild canid pack relies on the males to hunt and protect, while the females raise the pups and keep domestic order. For this reason, male wolves and their domestic cousins are generally larger than the females. This evolutionarily sound strategy might also account for the fact that male dogs of any breed are usually more aggressive, more apt to develop behavioral and territorial problems, and more prone to attempts at dominance among your family members. While male dogs are more active and playful, they may also be more destructive and unpredictable around children. Male dogs are more likely than females to mark their territory with urine and are ready to mate year-round. Male dogs that are sterilized, however, show a reduction in aggression, urine marking, and "mounting" behavior; they are also less prone to prostate infections and testicular cancer.

Female dogs are easier to train and housebreak before puberty, and are the preferred choice for novice dog owners. Many people have female dogs spayed—sterilized—because of their heat cycle: While wolves and some other wild canids have just one cycle per year, domestic dogs have two or more, unless spayed. Sterilizing females not only prevents the cycles, and thus pregnancies, but also reduces the chances of developing mammary and uterine cancers, depending on the sterilization procedure.

Contrary to popular belief, there is no gender difference for excitability, nervousness, or defensive barking habits.

COAT TYPES

While the survival of wolves and canids in the wild depended in part on the density of their thick, double-layer coat in cold weather and the ability to shed and allow good air circulation through their coat in warmer times, these factors are almost moot as survival tactics in domestic dogs. Similarly, wild canids needed to blend into their surroundings to hunt prey and elude predators, but coat types and colors are now more related to esthetics than survival.

While most dogs shed, the discarded hairs of some types of fur are more noticeable and omnipresent. Allergy-sufferers might want to avoid the heavy-shedding German shepherd, for instance, and stick with a poodle—this breed's fine hair is not shed, but instead must be trimmed by a groomer several times a year.

Along with their color, dogs' coats vary in length, type, texture, and density, from the herding puli's long, dense, water-resistant double coat to the stiff, short, bristly fur of the Chinese shar-pei. Make sure you are aware of any special grooming needs of the breeds that interest you. Also take into account the expense of having a dog clipped regularly. Breeds such as the poodle, Airedale terrier, and bichon frise need regular clipping. Call groomers in your area for sample prices and an idea of how often this needs to be done. (Turn to page 140 for tips on finding a reputable groomer.) You can buy a clipper and learn to do the job yourself, or you may want to consider another breed.

MORE THAN ONE

Thinking of bringing a second dog home? It may not be as easy as plunking the puppy down next to your loyal family dog. While dogs are social, they can have personality conflicts just as humans can. Introduce the dogs by the procedure described on page 121 before you leave them alone.

The new dog will have to adapt to the pre-existing family group, so choose a dog carefully in terms of temperament, age, and sex. Sterilized dogs of the same sex and calm disposition are fine, but choose a dog of the opposite sex if your first dog is somewhat high-strung or the least bit temperamental.

Your dog will probably adapt more easily to a puppy: He may take on a protective role and feel no threat to his established position in the family. But be aware that a non- or poorly socialized adult dog can influence a puppy's behavior and adversely affect his adjustment. A mischievous dog can also be a bad influence on an impressionable puppy. If you haven't yet learned to control your adult dog, you just may end up with two problem pooches instead of one.

Dogs and puppies who are well socialized accept other dogs more quickly—yet another reason to choose pooches who have had the necessary littermate bonding and socialization time when they were puppies. As a general rule, you should wait until your first dog is at least a year old and well trained before introducing another. If both dogs require obedience training, do this separately.

KIDS AND DOGS

Children and dogs make potentially blissful partners, but choosing a dog who actually likes to play with children is your best bet in orchestrating a match made in heaven.

Generally, sturdier, active dogs love kids, but unless your dog is well trained, he may pose a risk to your children. Smaller dogs may be hurt by unintentionally rough children, especially those under the age of three, but a well-trained smaller dog can be a great companion for gentler kids.

For best results, choose from among kid-friendly breeds such as the golden and Labrador retriever, boxer, beagle, West Highland white terrier, and Shetland sheepdog. A vet or dog breeder can suggest other possibilities.

Everyone needs friends, even your dog. But don't bring more than one dog into your home without considering a few important things, such as the dogs' compatibility with each other and with other pets, and the increased expense required on your part.

SHOPPING
· AROUND ·

Who is that doggy in the window? He could be peering out at you from a shelter cage, a breeder's fence, a pet-store display, or even trotting alongside you on the street, looking for a handout. There are many places to find a dog or puppy—some better than others. And there are many types and sizes of dogs in need of a good home. The trick is finding a dog who is healthy, happy, and as well adjusted as can be expected for an eager adoption candidate. Where will you find your dog or puppy?

PET SHOPS

Many people stroll through malls, glance up at just the right moment, and fall in love with a cat, bird, dog, or even an iguana in the pet-store window. While pet shops are a great place to buy food, toys, and other pet-related essentials, they are not your best bet for a four-legged friend.

Many pet stores do get their puppy stock from reputable breeders, but there are stores that buy their furry inventory from puppy mills. Breeding dogs en masse for profit (and not much else), puppy mills usually keep dogs in vile, inhumane conditions, with cages crammed together, no socialization of dogs with each other or with humans, and poor cleaning and feeding conditions. These circumstances do not produce well-adjusted dogs. Puppy mill dogs are often in poor health and tend to be nervous, distrustful, hard to train, and may develop behavior problems that can never be overcome, even by the most patient, loving owner.

Pet stores tend to regard dogs as "merchandise," but these outfits don't always have a return policy or other guarantees in case you have taken home an unhealthy dog. Some stores, however, have arrangements with local animal shelters to display and find homes for shelter dogs. The care the animals receive in these stores is comparable to that of a shelter—generally higher than in most pet shops—and in return for their effort, the stores benefit from selling food, toys, and other essentials to the adoptive family. Before you buy your puppy from a pet store, ask the store manager who supplies the animals—and insist on documented proof.

BREEDERS

If you have your eye on a specific breed of dog, and you've done the proper research to ensure that you can handle all its behavioral and physical tendencies—including shedding and exercise needs—your best bet would be

They may seem cute and cuddly, but look more closely at pet shop puppies before you fall in love with them. Often, their cramped living conditions and lack of regular human contact can result in health and behavioral problems. Possible signs of trouble include dull and patchy fur, listless eyes, and a dry, warm nose.

Opposite: This mixed collie looks well cared for in his shelter cage, but don't rely on appearances alone. As with breeders, stores, and other sources, make sure that the shelter from which you choose your new dog is reputable. A veterinarian will probably be able to provide this information. In most cases, shelter staff are friendly and informative, and can give you insight into the behavior and temperament of the dogs and puppies in their care.

A good breeder will provide the pups with comfortable, warm living conditions where they can interact with their canine family as well as with their human hosts.

to visit a breeder. Some people feel that dog breeders are a big part of the pet overpopulation problem, in that a breeder's overstock just ends up in shelters or pet stores. However, reputable breeders are very concerned with the welfare of their litters. They will try to find good homes for the puppies, and sometimes even raise unsold dogs themselves.

Caring breeders usually require that buyers sign an official contract. It helps to educate the new owner as to the puppy's needs, and it typically includes clauses that forbid further dog breeding without the breeder's permission; forbid the puppy's sale, abandonment, or transfer of ownership; ensure the puppy's spaying or neutering; and offer a return or refund if the pup either develops a hereditary illness or disease within the first year, or if the new owner can no longer care for him. The American Kennel Club (AKC), your veterinarian, or the local animal shelter can provide a list of reputable breeders (page 210).

Before you visit any breeder, it's a good idea to prepare yourself with a list of questions. For example, with breeds that are prone to hip dysplasia—such as the Bernese mountain dog, German shepherd dog, golden retriever, Rottweiler, or St. Bernard—ask if the puppy's parents are certified by the Orthopedic Foundation for Animals (OFA) or how well they scored on the hip dysplasia evaluation of the PennHip (University of Pennsylvania hip improvement program). Ask if the dogs have been socialized—with dogs, humans, and the normal sounds and sights of home life. A conscientious breeder will provide pups with good living conditions where they can interact. If possible, arrange to meet the litters' parents—the breeder ideally should keep the mother or both parents with the litter until they are past the weaning stage and are well socialized. If the parents are not well-adjusted, their behavior and attitudes may have already been imprinted on the impressionable puppies

When visiting the breeder for the first time, check to make sure that the dogs' living conditions are clean and well-maintained. When you spot a pup you like, do a quick test to see if his temperament matches what you're looking for. Overt aggressiveness or meekness are red flags. (See page 81 for more on temperament and health checks.)

You should expect an inspection yourself. Remember that upstanding breeders will have a discerning eye as well—the pups are more than products and profits to them—so be prepared to answer some questions about yourself and your intended commitment to your new family member. Breeders should be able to supply you with a good deal of information regarding the proper care and upbringing of your puppy.

Any reputable breeder will have official, documented proof of the puppies' immunization and pedigree records. Ask to see them. You'll also get a copy to bring home with the pup. Take the immunization record to your veterinarian so she can start a medical file right away.

SHELTERS AND HUMANE SOCIETIES

A shelter or your local humane society can be good places to find a dog to suit your needs. (For tips on finding one, turn to page 210.) The range of choices will be large—young dog or mature, furry or sleek, big or small, or any of the other criteria that are important to you. But it's best to visit a shelter only after you have a good sense of what you are looking for.

There's no mistaking a healthy dog, such as this Shetland sheepdog (Sheltie) above. A shiny coat, bright eyes, pink gums, white teeth, and wet nose are obvious indications. But don't let appearances rule: Do a few simple tests to check whether the dog that you have your eye on will be a treasure or a terror (see page 81).

LEMON LAWS

Cars, toasters, and even clothes have return policies, so why not dogs? In many places, officials recognize that pet owners need to be protected from unscrupulous sellers as much as any other buyers.

In New York state, for instance, pets are guaranteed for health and veterinary treatment—mostly within the first two weeks of purchase. If a vet finds your new puppy or dog is unfit due to hereditary or congenital illness, or infectious disease, commercial and private pet dealers must either allow a return and refund of all costs, a return and pet exchange, or they must pay all medical treatment expenses.

Your state attorney general's office can apprise you of any recourse available to pet owners when *caveat emptor* hasn't been enough.

There's no such thing as an ugly puppy, but some are more suitable as pets than others. When choosing a dog from a litter, check for signs of health and character. Aggressive dogs and those who are too meek are best avoided. Of course, if you're willing to put in the effort, any healthy dog can provide years of love.

Many dogs come with a history, and if they know it, the shelter staff will be happy to share with you. For example, if they know a dog has been abandoned or abused, they can pretty much tell you what you are getting and explain the behavioral problems you may have to overcome. Ask the staff enough about the dog to make sure that he will fit well into your life. If you have other pets at home, ask if the new dog has ever lived with other animals and how he is doing in the shelter. If you have youngsters, take them with you to help choose the dog.

Finally, check that the dog you are bringing home is in good health and has had his shots on time. There is nothing more heartbreaking than bringing home a puppy from a shelter, only to discover soon afterward that he has come down with a preventable but fatal disease such as distemper or parvovirus. Also, check into the possible breed-related conditions, such as the skin allergies of the West Highland white terriers, to which your dog may be prone.

Don't rush through the selection process. It will take an hour, maybe more, for you to cover all the bases. Hold off on falling in love with your dog until you're sure that everything has been checked and double-checked, and that you are right for one another.

THE VETERINARIAN CONNECTION

With their many clients, and their associations with breeders' groups, kennels, and shelters, veterinarians are very well connected in the dog-adoption community. If a client's dog gives birth to a litter of puppies, if someone is looking to find a good home for a dog they can no longer keep, or if there's a friendly stray in the neighborhood, a vet is usually one of the first people to get wind of it. Bulletin boards in veterinarians' offices can be an excellent source of information if you want to bring a dog into the family.

A vet can also help you select a dog—matching your needs and expectations with the known behavioral and health issues of any breeds you're considering. Make sure that the vet examines your new pet for physical and emotional problems as soon as possible.

OTHER SOURCES

Other easy ways to find a dog are through friends or your newspaper. Sometimes the best dogs are previously owned—they're trained and mature, and their medical history is fully available to you. Just make sure that the dog you choose was not given up because of any hidden illness or behavioral problem.

Many of the puppies advertised in newspapers are the result of "accidents": the parent dogs were never spayed or neutered. Provided you take the same care as you would when choosing a dog in any other situation, you can adopt one of these pups, but make sure that the litter was an innocent accident, and not a calculated move to make a quick buck by selling puppies. People mainly interested in the money cannot always be relied upon to take good care of the puppies in the critical first six to eight weeks of life, much less to see to all the needs of the pregnant mother dog.

Dogs who were chosen to be seeing-eye dogs or other special-assistance dogs but didn't make the final cut can be excellent pets. They were initially singled out for the program because of their calm, gentle, intelligent nature, and, as a bonus, have been fully obedience-trained.

HEALTHY AND HAPPY DOGS

A dog is a dog is a dog? Not really. A dog who is ill, diseased, or emotionally withdrawn is not likely to provide the years of happy companionship that every dog lover wants. That's why you must give the dog or puppy you are considering for adoption a thorough once-over.

Look at all the pups in the litter, or those in the shelter. Pass on those who show signs of listlessness, extreme fear, and overt aggression. Of course, landing in a shelter can be unsettling for any dog—especially a longtime pet—so some fear is understandable. Pooches who come right up to you, give a few curious sniffs, then continue exploring their surroundings are your best bets. Cowering, snarling, and barking may be signs of a nervous or ill dog—more trouble than the average person is usually prepared to handle. Look beyond your own sympathy to find the alert and friendly dog with bright eyes and a healthy glow, one who prompts a smile when you look at him.

TAKING IN A STRAY

Every year, thousands of dogs are abandoned or born from unplanned litters. Some will be turned in to shelters, but many will die of starvation, accidents, or abuse. While it is noble to take in a stray dog, you should know beforehand what you're getting yourself into.

First, make sure the dog is not just lost. Post flyers, check the lost and found section of local and community papers, and call shelters to see if any dogs have been reported missing. Look for a tattoo on the dog's inner thigh—this is either a breeder's marking, which can help identify the dog's origin, or an identification number linked to a national registry that would have the owner's name on file. Your vet may be able to help. Also ask your vet or the shelter if they can scan the dog for an identifying microchip.

Once you're assured that the dog is safe to adopt, take him to the vet for a full check-up. Isolate him from other pets until you're sure he's fine. You can even start his training during this time. Once everything checks out, you can take the pooch into your home—and your heart.

Stray dogs should be approached carefully, then fully checked out by a veterinarian— and even a trainer—before you open your home to them.

DOG-PROOFING
·YOUR HOME·

Temptation is a full kitchen garbage can. Food, paper napkins, and assorted other bits can keep your dog coming back for more, no matter how many times you scold him. Your best recourse is to hide this tasty—and sometimes hazardous—challenge, or use a garbage container with a locking lid.

Crash! If your dog is not accustomed to your home, or if you have yet to train him, get used to this sound. A dog who cannot be controlled is like the proverbial bull in the china shop, wreaking havoc in your house. For his safety and for yours, it's imperative that you go to every possible length both to train him *(page 100)* and to dog-proof your home.

Wolves and other wild canids rely on their natural curiosity to seek out food and shelter, and to keep watch for enemies and other dangers. Nowadays, this evolutionary holdover is seldom linked to survival, but the inherent tendencies must be anticipated by owners. Puppies, in particular, inadvertently find themselves in a world of trouble when in a new environment. Even at home, surrounded by familiar people and items, it is in a dog's nature to look for mischief and adventure.

If you haven't taken the time to prepare your house for your new arrival, your belongings may pay the price. Don't forget to secure your yard, as well. Whether you want to keep neighbors safe from your dog, your dog safe from the dangers of the street, or a well-groomed lawn safe from a digger, plan to dog-proof your property even before you bring your pooch home.

POKING AROUND
Preparing for a dog is like preparing for a baby. Crawling around your house will alert you to the dangers that lurk for any being who lives only a foot or two from the ground. Make a note of the trouble spots your dog will invariably find and want to explore. Electrical outlets, for instance, are one of an inquisitive pup's worst enemies. Even older dogs that have been trained to keep their distance may sometimes sniff at or lick these interesting items. The best solution to this potentially fatal problem is simply to cover the outlet. Hardware stores sell childproof outlet covers; your pet store may carry similar items.

Similarly, cracks in the wall, gaps between the wall and baseboard, and gaps between loose floorboards are irresistible to a curious puppy. Before you let your dog loose, call in a carpenter, or fix these things yourself. Do a thorough job—even slight problems will be magnified by incessant pawing, nose-poking, and gnawing. Remember to clean up after any home repair projects; the materials and products used may be dangerous to your dog (see page 206 for a list of poisonous substances).

ROOM-BY-ROOM

The kitchen, bedroom, bathroom, and even family room hold a world of mystery and temptation for any dog—there's always something to do. But while gnawing on the sofa cushions is a relatively harmless canine pursuit, despite the sure scolding from his human companions, chewing on or otherwise investigating other household objects may prove to be deadly.

Kitchens are especially dangerous. They are rife with sharp cooking and cleaning utensils, poisonous detergents, sharp aluminum cans, and choking and suffocation hazards such as discarded chicken bones and plastic bags. Also, a shattered glass or dish can add up to a dangerous situation for an animal; cleaning up the shards thoroughly is a must. Lock cabinets where plastic bags and detergents are stored, keep cutlery and dishes away from a dog's access, hide garbage bags out of sight, and be vigilant about cleaning up messes. Teach your dog to keep out of the kitchen—or any room—at your command to allow you the time to clean up breakages and spills.

Family rooms, dining rooms, and bedrooms all hold the allure—and the danger—of loose electrical, curtain, and blind cords. Remember, anything that should be kept from a child's hands should also be bundled, stored, or tied beyond a dog's access. You can use cayenne pepper spray or any bitter-tasting or foul-smelling repellent available at pet stores to make certain

Your dog's chewing and other destructive behaviors don't have to rule your life. Take your dog to obedience class, reprimand him for chewed shoes and other damaged property, and never buy him toys that resemble items you want him to leave alone.

Puppies and dogs love to dig, chew, and explore. Cut flowers and fragrant houseplants are a favorite target. While it's best to keep greenery inaccessible to a dog, you can't always prevent trouble. To be safe, make sure that the plants and flowers you keep are not harmful to your dog.

areas less than tempting for your dog—for instance, if you want to train him to stay off couches and beds. Teething puppies may chew almost anything to ease their discomfort, but a nylon chew-toy or wet, knotted towel left overnight in the freezer can provide soothing relief. Never buy toys and treats in the shapes of shoes or clothes; that would only encourage dogs to go for the real things. Don't use old clothes and shoes as toys, either, since dogs can't tell the difference between old and new. Latex, nylon, hard plastic, and rawhide chew toys and bones will do your dog just fine.

In bathrooms, keep all medications, air fresheners, and personal-care products out of a dog's way. Most dogs love to slurp a little cold water from the toilet. While this privilege may not bother some dog owners, be aware that it can pose a danger to smaller dogs that may fall in as they try to hoist themselves up for a drink. Keep the lid down when you aren't around to supervise. Don't forget that most toilet tank fresheners are poisonous. Remove them, find a pet-friendly product, or keep the lid closed altogether.

Finally, be extra vigilant when it comes to the garage. Paint, chemicals, motor oil, and sharp tools add up to one thing: This is no place for a dog.

IT'S NOT EASY BEING GREEN

Plants may be a tempting, tasty treat for dogs, but chewed vegetation is the bane of any proud gardener's existence. Worse, many indoor and outdoor plants are toxic to dogs. Either make sure that your houseplants are not harmful, or keep them high up, out of reach. Garden plants to avoid include potatoes and spring bulbs; among the indoor hazards are ivy and azaleas. See page 207 for a list of plants that are unsafe for your dog.

When walking outdoors, keep your dog on a leash and close to you to more easily monitor what he's getting into. Avoid areas that have been sprayed with insecticides or other poisons. (Lawn-care companies usually plant small warning flags in freshly treated areas.)

DON'T FENCE ME IN

It may arouse your sympathy to see a dog locked behind a fence instead of playing with neighborhood children. But far more tragic is a dog who has been hit by a car because he has been allowed to run loose in the street, or a dog who must be destroyed because he bit someone. Fences and gates are not cruel—in fact, they help reinforce a dog's natural protective tendency. Your home is his home, and this extends to the yard as well. While wild canids have a general sense of the territory that their pack inhabits, a fence can help the domestic dog to more concretely define home base. Territoriality applies to the dog crate or cage, as well, and your dog will extend this territoriality to the room in which the crate is placed, then further extend his domain to include your house and yard. Introducing your dog to a crate or cage is of benefit to your home and belongings *(page 105)*. A crate offers a secure and familiar setting—just what a dog needs—and unless he suffers from separation anxiety, he can happily stay in the crate for a few hours while you go out.

Before you allow your dog free reign of the yard, make sure that your lawn is safe and dogproof. Many dogs—and their wild cousins—love to dig. This is not just a fun pastime for them. Their ancestors (and some domesticated dogs today, especially terriers) bored holes in the earth to catch rodents or other prey whose scent they picked up. In addition, canids are genetically programmed to seek cool shelter on hot days; a well-dug hole provides this. In the wild, pregnant canids have always dug holes in which to hide their defenseless pups from predators and the elements. To keep your lawn beautiful, and your vegetable garden safe, you may want to spray the area with a store-bought, plant-safe pet repellent. Finally, train your dog to dig in a specified area only (page 117).

A fence around your yard not only keeps passersby safe, it also protects your furry friends from harm. The gate should be sturdy, high enough to keep jumping dogs in, low enough to the ground to keep small or digging dogs secure, and it should have a lock. You may want to try one of the newer electronic "invisible fences": Sensors are buried at the borders of the lawn, and a special battery-powered collar around your dog's neck emits a slight shock when he approaches the pre-determined boundary. It's relatively painless, and your dog quickly learns when he's gone too far.

FIRST
·DAYS·

A puppy—here, a Jack Russell terrier—will get used to new surroundings more easily if allowed to explore. But keep him on a leash for the first few expeditions. A quick tug will teach him that you're the boss, and your guidance will ensure that he stays out of trouble and harm.

When you get a new dog or puppy, you know he will be an important part of your family, but during the first few days, make no mistake, he'll be your primary focus. Before you can take him on long walks or scratch his ears as the two of you watch television, there will be a period of adjustment—for both of you. To ease the transition into your family, treat him with extra care during his first days in your home.

SETTING UP BEFOREHAND

Before you head out to pick up your new four-footed family member, you need to prepare. First, buy a flat collar and leash, and make sure you have those with you when you go to get the dog. A puppy's neck will grow— and grow fast—so be prepared to replace the collar a few times. Fit the collar *(page 104)* so that there is just enough room for one or two fingers to slip between the collar and neck. Flat collars may be nylon, leather, or some other sturdy material, but don't purchase a very thin one. It may dig into the dog's neck when you tug on the leash. Choke collars may be used for training *(page 102)* but don't use a choke as your dog's permanent collar. The movable end ring can get caught on anything, tightening the collar around the dog's neck as he struggles against it. Or the chain itself can get caught on the dog's lower jaw—an uncomfortable and frightening experience for your pooch.

Proper identification tags are the next step—make sure they indicate your name, address, and phone number. (For more information on ID tags and other identification for your dog, turn to page 209.) Later, you'll need license tags from the city—if required—as well as veterinary tags indicating that your dog has been properly vaccinated.

Food and water dishes—plastic or metal—should be deep enough to hold adequate nourishment, but not so deep that your dog's ears droop into the bowl. No matter where you get your dog or puppy, ask about his diet and continue feeding the same thing for a little while. Your dog will be nervous after the big move, and possible stomach troubles from an abrupt change to new food will only add to his distress. After your dog has settled in a bit, your vet may recommend gradually switching to a different food. Puppies are usually put on special growth food until they reach maturity. An older dog may need a special diet if he has a medical problem or a tendency to be overweight.

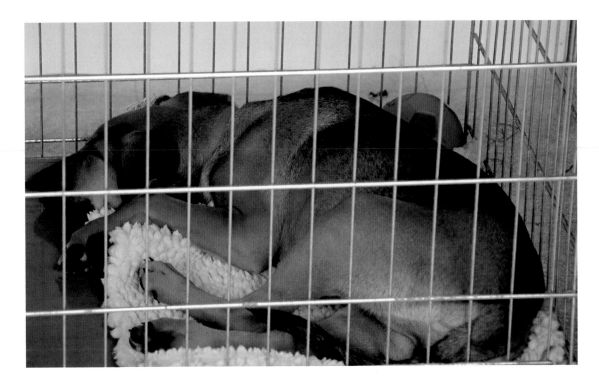

Far from causing harm or distress, a cage or crate will give your puppy comfort and a place to call home. Buy one that will be large enough for your dog's full adult standing size, and block off part of it with a plywood barrier or cardboard boxes while the pup is small. Make a cozy sleeping place using towels or blankets, and keep a few favorite toys inside. Move the barrier to allow more room as your dog grows.

A WHOLE NEW WORLD

Even the most outgoing pup will need a period of adjustment before he feels secure again. If your new dog is a little quieter at first than when you chose him, just give him time. He'll come around.

Dogs need to explore their environment in order to gain a sense of security. Your new dog will want to check out every part of his new home: let him. But watch him closely to keep him out of trouble. It's best to leave the dog on his leash, and either let it drag behind him—with you nearby—or take it loosely and follow along on his journey of discovery. Your companionship on this first daunting expedition will tell your dog that you're there for him, and also remind him that you are in charge. A gentle tug on his leash when he gets too near the stove, garage, or other danger will reinforce this.

It's normal for a nervous dog or puppy to wet the floor. Take the new arrival to his designated elimination spot, preferably outdoors, as soon as you bring him home. After he has eliminated, make a big deal about praising him; this will start the positive reinforcement necessary for toilet training *(page 105)* right away. If he has an accident before getting to his special area, don't yell—this will only cause further stress, and it won't do much for the bonding process, either.

Don't invite all your friends over to meet the puppy right away. It's important that he form ties with your family members first, and learn his place in the pecking order. Too many people at once will both scare and confuse him, neither of which helps develop a well-adjusted pet.

Like wolves, dogs are social animals that like to eat together in a group. For this reason, it's a good idea to place your dog's food and water dishes where he can at least see his human family. Of course, it's up to you to prevent the habit of begging for food. (See page 119 for more on mealtime training.) Bring your dog to the dining area for the first few days, fill his bowls, then set them down. When he starts to eat, convey praise. He'll soon learn where to go at mealtime.

If you already have another pet at home, you'll have to introduce him to the new addition. Don't push for the two to become friends right away. They must feel each other out and establish trust and respect—just as a first child might with a newborn baby. Keep a close watch on the two, and never leave a puppy alone with an adult cat or dog; if the older pet feels his place in the family is threatened in any way, he might just take it out on the new arrival. Expect the older pet to exert his special status a little. He may taste the new arrival's food, use some of the new toys, and maybe even check out the newcomer's sleeping quarters. Don't ignore your older dog in favor of the new one—train him to respect the new kid on the block. (See page 121 for more on bringing a new pup or dog into an already pet-friendly home.)

THE NAME GAME

If you want your new dog to come when you call, he'll need to know what to answer to. The numerous incidences of "Rover" and "Sparky" notwithstanding, naming your dog can be a challenge. A true animal lover will try to match the name to the dog—either his personality, looks, or behavior. Others name their dogs as they would babies—after relatives, their favorite places, or even public personalities. Remember that it's easier to train a dog if the name is not too complicated. A one-syllable name might be confused with other words, but keep it to no more than three syllables. Once you've decided upon a name, use it often, and tell others to do the same. Only begin using nicknames when you're sure your dog knows his real name.

FIRST NIGHT

Removed from his home and thrust into a new family, your dog might miss his old, familiar surroundings, especially if he was taken from his litter. Many people assume that their dog will sleep at the foot of their bed, or on the floor in a warm room. This might be the case, but only after the dog has settled in and feels secure. To help ease the transition, get a crate—wire or plastic—or a comfortable dog bed or pillow. Crates can also be used for house-training (page 105). All these items are available at most pet-supply stores. Some crates are solid, some are collapsible, and some are better suited for traveling with your dog. Check with the pet store or your vet for advice on the best size and type for your needs. When you bring it home, line the crate with a warm, washable blanket, drape a portion of it with another fabric for privacy, and

keep some toys inside. Dog beds and pillows should be washable, and large enough to let your pooch stretch out. Always remove your dog's collar before he enters the crate, and never leave him inside for more than a few hours (*page 105*)—unless it's overnight and you're there. Place the crate in a corner of the family room, or some other quiet area where the dog can feel like he's part of the family. As an added bonus, your dog's natural instinct to keep his sleeping spot clean will help ensure that there won't be any messes.

For the first night, some people keep a ticking clock near their dog's sleeping area to approximate the sound of his mother's and littermates' heartbeats. Others leave a radio on so the dog won't feel alone. If your new family member cries or whines continuously, check on him. He might just need the reassurance that he's not alone.

DOGS AND KIDS

Whether it's their first exposure to dogs or not, kids need to be extra gentle around a new dog. Rushing him, playing with his toys before he's secure, and being too loud will only scare him and may make him defensive—possibly, dangerous. Supervise your kids' time with the new dog at first, and teach them to play gently. Also, show them the right way to pick the dog up: one arm around the chest, the other supporting the rear. Small dogs, especially, need to be handled with extra care. Larger dogs must be trained to be gentle themselves, lest they get too rough with the youngsters during playtime.

With a shared love of play and adventure, dogs and kids make fast friends—as long as they respect one another. Make sure children understand how important it is to let their new dog have the space he needs—especially at meals and nap-time. Remember, an aggressive, uncontrollable dog is even less desirable when kids are added to the mix. Both dogs and children need training.

IN THE
COMPANY
OF
DOGS

· · ·

**"You think dogs will not be in heaven?
I tell you, they will be there
long before any of us."**

·

ROBERT LOUIS STEVENSON

READING

·YOUR DOG·

Learning to read a dog is much like studying a foreign language. It requires concentration and recognition that communications can have quite different meanings in different cultures and depend on the context within which they are sent. Dogs bark, whine, and growl, but mostly they "speak" via a body language designed to be understood dog-to-dog. The meaning may not be intuitively obvious to humans; decoding requires some practice. To understand "dog" successfully, we must stretch beyond ourselves into canine culture. What are the rewards of making this effort? Clearly a better reading of unknown dogs can prevent the occasional nip. But far more importantly, we gain the chance to interact more fully with the dogs who share our lives. We can reach more of an "inter-species" understanding and deepen our bond by our ability to communicate. It seems only fair. Over the centuries, dogs have become very adept at interpreting human body language and even at learning spoken words and hand signals. Now it's our turn.

COMMUNICATING SOUNDLESSLY

A dog's tail, ears, eyes, and mouth speak volumes without making a sound. Everybody recognizes a rapidly wagging tail as a sign of canine excitement, but the tail also is a primary conveyor of social standing and mental state. Don't make the mistake of automatically interpreting tail wagging to mean friendliness. Generally, a tail held above and away from the body or curled over the back denotes dominance and, especially if accompanied by bristling of the hair, threatens aggression. However, some dogs, such as the Siberian husky, have tails that curl up naturally, and would appear perpetually dominant based solely on a tail reading. A relaxed dog, comfortable in its surroundings, generally holds its tail lower and away from its body. On the other hand, a frightened or submissive dog holds its tail close to its body, tucked between its legs. But be aware that some breeds—greyhounds and whippets, for instance—naturally carry their tails between their legs, whether submissive or not.

The secret to deciphering ear communication is knowing how the dog carries its ears when relaxed. When the ear is more erect and pulled slightly forward, the dog is attentive. Combined with a head tilt or slightly open mouth, this shows interest or an attempt to understand. A frightened dog, or one protecting against attack, pulls the ears back and flat against its head.

A domestic mixed-breed *(above)* mirrors the aggressive posture of his gray wolf counterpart *(left).* Whether displayed by a small domestic dog or a wolf in the wild, this body language means business. Note the similarity in appearance: In each, the hackles are raised; the lips are pulled back in a snarl showing the large canine teeth; and the stare is intent on the subject of aggression. You can almost hear the growl. Few canines, or humans either, would misread the signals: "I am ready to bite!"

Overleaf: Longhaired dachshunds

These two pups hold each other's eyes, but the message is a positive one. With tail held up at a jaunty angle, elbows down flat on the ground, and bottom lifted high, the dog on the right displays the typical position preceding frolic, the play bow. The canine buddies share only one thought, "I want to play!" If these two bark, it would be out of excitement, not threat.

Of course, the controversial cosmetic surgery to create perpetually upright ears (*page 174*) makes ear reading more difficult.

Eye communications range from the confident stare of the dominant dog, which can harden and intensify if the dog feels challenged, to the averted, downcast eyes of the submissive dog. Although dogs' mouths are less expressive than ours, they still send a message to the attentive owner. A dog with a relaxed and slightly open mouth is happy and comfortable. Yawning, on the other hand, is more often a sign of anxiety than weariness or boredom. So those yawning pup patients in the veterinarian's waiting room are probably not so much world weary as apprehensive about their impending visit. If its lips are curled enough to show major teeth, skin is wrinkled above the nose, and its mouth is partially open, the message is a clear warning: the dog might bite.

LIVING THE LANGUAGE

A dominant dog walks on its toes, often leaning forward, with a stiff gait. Ears and tail are up, the head is high, and the dog meets your gaze confidently. If it senses a challenge, its hackles rise and it stares more intensely. Your return stare, regardless of how sincerely and kindly meant, may be seen as a challenge and could elicit a bite. When meeting a more submissive dog, the dominant dog may attempt to place its muzzle or paws across the subordinate's shoulders or back. If a dog is highly dominant, it may respond to your touch on or at the back of its head with a growl or snap, reading into your hand position an attempt to express your dominance.

When they're feeling playful, dogs assume the easily recognized "play bow" *(opposite)*, with tail up, front legs on the ground and an expectant, alert look. The dog may bark, but the context shows it is an excited, not a threatening, bark. An interested dog also exhibits this alert look, standing with mouth partially open, often with his head cocked to one side *(right)*.

The fearful dog recoils, its ears flat and tail tucked, but it may also show signs of aggression with raised hackles and teeth bared. When confronted with mixed signals like these, always heed the ones from the "sharp end." This dog could bite, although out of fear, not to show dominance. The submissive dog crouches down with its ears back, eyes averted, tail low or between its legs. In a more extreme submissive display, the dog gradually rolls over onto its back, exposing the belly. The animal may even urinate a few drops, perhaps a throwback response to the first authority figure in its life, its mother, who stimulated her pups to urinate and then cleaned them up. Submissive urination is easily misunder-stood, especially if produced in response to the owner's anger over some infraction.

This Australian shepherd shows the typical look of the alert dog. His face, with mouth partially open, shows an inquisitive, interested expression and his whole body position reflects the expectation that something exciting is about to happen.

From a human perspective, the dog may seem defiant, even spiteful. But far from committing an act of defiance, this dog is trying to placate the angry owner by showing extreme submission.

HARK! HARK! THE DOGS DO BARK

While dogs' primary communication is via body posture and position, they also do some vocalizing *(page 57)*. Many dogs seem to enjoy a good bark—especially combined with howling—often to their owners' frustra-tion. A bark can express many things, from sheer joy at the thought of a game of ball to celebrating your arrival home or warning of an intruder. When a gentle bark accompanies a nosing of the leash or a tentative paw on your lap, it may even be a question or suggestion. Dogs will also growl when threatened, whimper and whine when seeking attention, and yelp in fear or pain. In each of these situations, a combination of the dog's body language and an understanding of context are vital to understanding your dog's message.

BONDING WITH
· YOUR DOG ·

Living with a dog is a winning situation for both humans and canines. Pet owners enjoy lowered heart and pulse rates and reduced cholesterol and triglyceride levels. Dogs offer us emotional support, lift our spirits, and ease stress. In turn, we provide affection, leadership, and physical care for them. But what is the basis of this dog-human bond and how can we strengthen it?

THE GROUP PERSPECTIVE

Canids are essentially social animals; in the wild, their survival depends on the pack. It is this pack mentality that motivates the wolf to identify a clear leader (or become that leader) and to fill a position within an established hierarchy. As a pack member, a wolf has a powerful attachment to the lead or alpha wolf, as well as a strong loyalty to fellow members and an instinctive desire to defend the pack's territory. When it comes to domestic dogs, this concept of social order is particularly well-suited to relating to people. In general, dogs are quite content to live in a human family group, accept us as leaders, and act cooperatively with their human family members. They are highly motivated to work with people, whether in a direct protective role as police dogs, in the guiding help of assistance dogs, or in the supportive role of therapy dog or beloved family pet.

In order to nurture and strengthen these instinctive drives, we first need to recognize that dogs are a separate species with unique needs that deserve respect and understanding. They want—and need—to know where they belong in the family. But more than knowing their place, dogs need to learn the rules of the household. We owe them strong, consistent leadership; every family member needs to make clear what is acceptable and what is not. Dogs are motivated to please their leader but don't instinctively understand what humans want. It is our responsibility to uphold our side of the dog-human equation by clarifying our expectations and making sure they are reasonable, given the dog's age, breed, and personality.

ESTABLISHING AND STRENGTHENING THE BOND

Establishing a bond is easiest and most natural with puppies. Leaving their mother's authority at about eight weeks (taking puppies from their litter earlier than six weeks means the puppy misses an important part of the "litter" experience and may always have difficulty relating to other dogs),

These two friends share a quiet moment, conveying their good feelings for each other. The dog is perfectly comfortable to share close quarters with his human buddy: Like their wild counterparts, domestic dogs enjoy physically touching and being close to those they trust and depend upon.

Opposite: Much like the welcome home our domestic dogs give us, wild dogs greet each other with affection, shown here by body contact, licking, and mouthing. Such grooming behaviors reinforce the bonds that already bind the pack. Wild canids depend on their fellow pack members for survival as they live, hunt, and raise young communally.

With his master holding his leash, an Australian shepherd goes confidently over an A-frame obstacle. During agility work, dogs have fun, build self-confidence, strengthen muscles, and increase flexibility by going over various hurdles and other obstacles, traveling through tunnels made of plastic sheeting, and threading around closely set poles. Joining a local agility group and meeting for practices and competitions will expand the social life of both you and your dog. Agility is open to all dogs, regardless of their parentage—mutts are welcome. By becoming involved in these challenging and fun physical activities, you and your favorite canine strengthen the dog-human bond.

the impressionable puppy willingly comes to see his human as the leader. You reinforce this natural tendency by calm, patient, consistent attention and allowing the puppy to adjust to new routines. The pup is anxious to trust and bond and will respond positively to your care. Provide him with a safe sanctuary (such as a crate) that is quiet and puppy-proof. Build his confidence with unambiguous expectations and gentle treatment. Young puppies are easily intimidated, so be careful when training. A firm "no" is usually sufficient when discipline is necessary.

Establishing a bond with an adult dog may take a little longer, especially if he comes from a troubled past, but it is well worth the effort. If possible, learn as much as you can about your adult dog's early life so you can avoid inadvertently repeating any troublesome or frightening experiences. Assert your leadership in a non-confrontational way. Be clear and firm in your expectations, and be patient.

Whether your pet is a puppy or adult dog, each activity the two of you share builds and strengthens your bond. Stroking, patting, and gentle grooming comfort and please your dog. Appropriate play allows you to share your dog's natural joy in movement and games. Time spent in agility training and obedience work enhances your relationship while allowing your dog to gain confidence as he successfully completes tasks and earns your approval as well as the occasional treat. Make sure that the sessions are no longer than the dog's attention span and that they always end on an up note with the dog successfully completing a command and winning praise.

Dogs understand and learn from the consequences that follow their actions. By rewarding good behavior with praise, attention, and honestly given rewards, you reinforce both the behavior and the bond between you and contribute to your dog's sense of confidence and his place in the family structure. If the dog's behavior is unacceptable, make this clear in a firm, negative tone of voice. Intimidation, cruelty, or any attempt to hurt or harm your dog betrays his trust, damages the bond you share and can lead him to be fearful.

Dogs come to us with a natural sense of order and a willingness to acknowledge humans as top dog. In response, we need to reinforce our position by confident and consistent leadership and guidance. Understanding your dog's world-view and being a firm and consistent leader will allow him to grow to his full canine potential and permit the bond between you to continue to develop.

A Great Pyrenees therapy dog visits a nursing home resident, bringing the comfort of a furry head to pat. Therapy dogs often reach patients otherwise cut off from the world by age or illness and provide the opportunity for residents to share happy memories of dogs they have known and loved. Therapy dogs need stable personalities and strong obedience training to qualify for their very fulfilling work. If you feel that you and your dog would enjoy volunteering at a nursing home or similar location, call an institution that participates in the therapy dog program, your local breed club, or an animal rescue organization for more information and training.

"AND BABY MAKES FOUR"

A baby brings tremendous changes in a household. Begin to accustom your dog to any expected changes in daily life a few weeks before the baby's arrival, so he will be comfortably settled into the new routine.

If your pup's been backsliding on training, start reviewing basic obedience. Play a tape of baby sounds and start using baby lotions and powders so the dog becomes used to baby noises and smells. After the baby arrives, let the dog smell some of the infant's clothes or blankets.

At the actual introduction, one person should hold the baby while another keeps the dog on a leash. Allow him to see and smell, but not touch the new arrival. The dog needs to become used to baby squeals and jerky movements. Feed, praise, and play with your dog in the baby's presence so he associates these pleasant experiences with the baby. Be very sensitive to warning signs of fear or aggression. Even if the dog seems to accept the baby, never leave an infant or young child alone with any dog, even the family pet.

BASIC
· TRAINING ·

The pup's formative early weeks are the best time to expose him to as many different experiences as possible. This puppy has already become comfortable with children, sandy beach surfaces, and the ocean. Of course, young animals, like young children, can be overwhelmed by too many new stimuli at once. Introduce new situations, then let the pup proceed at his own pace. Encourage him to climb stairs; take him on car rides—to places other than just the vet—so he learns to associate cars with fun; take him out on busy streets so he gets used to the hustle and bustle of traffic and people of all races, sizes, and manners of dress; and let him meet other dogs (after you ensure they are friendly). The more experiences and adventures the pup has at an early age, the more confident and reliable he will be, and the more likely to face future situations with interest, not fear.

A well-trained and socialized dog is a joy. But such paragons do not appear magically; they are the result of patient, consistent training and obedience work. A dog wants to know his place in the family hierarchy, and positive but directive training makes clear to the dog that you are his leader. By making our messages to our dogs straight and to the point, we show clear leadership, take advantage of the dog's sociability, and make living together in general (and training in particular) a lot easier.

TRAINING THE TRAINER—YOURSELF!

Regardless of whether you hire an individual trainer, attend obedience classes, or tackle the project on your own, you need to expend some effort, and possibly expense, on training your dog if you expect to live happily with him. For first-time dog owners, professional trainers can explain successful techniques, deal with problem areas, and help the owner form realistic expectations of the time and effort needed to guide a dog into a happy, obedient future. Even for an experienced dog person, an outside trainer can help sharpen skills and suggest different approaches. Attending a class gives your pet the additional benefit of opportunities for socialization. You can find a qualified trainer or obedience school through your veterinarian, the local humane society, or SPCA.

There are almost as many approaches to dog training as there are dog trainers. The various theories range from reward-only training to relying on physical punishment to elicit the desired behavior. What works best for most dogs is a strong emphasis on rewarding appropriate behavior and correcting the dog only as necessary. Correction may be a frown, a verbal reprimand, or a "time-out," where the owner withdraws his physical and emotional attention from the dog. Corrections should assert the owner's position as leader, but should never inflict pain on the dog. Give *yourself* a time-out if you find you are becoming angry at your dog or so frustrated that you cannot convey your expectations to him. Harsh or inappropriate corrections confuse and upset dogs, making training almost impossible. Physical punishment ultimately elicits negative responses, even aggression. If a dog "learns" anything from physical punishment, it is to fear its owner, be hand shy, and distrustful.

Most dogs are anxious to please. Often the block to effective training is not a recalcitrant dog but a confused one. Dogs do not instinctively under-

While this pup may look wistful, he is also likely comfortable and content in his crate. Early crating as part of puppy training accustoms him to having a quiet place of his own, where he can play with his toys or enjoy a snooze. Having a pup or dog crate-trained allows you to keep him (and your house) safe when you can't be right there supervising, although in some cases of separation anxiety, a crate may actually be dangerous *(page 115)*. A crate also serves as good travel accommodations and a safe place to recover after surgery or illness. It's best to crate-train in puppyhood, since it takes more patience and perseverance to train an older dog. An adult dog who was crate-trained as a pup will willingly spend time in his private room without complaint.

stand what actions and behaviors are valued by their humans. You need to convey commands in a form your dog can comprehend; this means training yourself as much as training the animal, so that you don't inadvertently reward misbehavior *(page 114)*. Dogs exist very much in the present tense. Because of this mind-set, timing is everything in terms of both rewards and corrections. The reward for obedience must immediately follow good behavior. Similarly, a correction must be directly connected with the undesired behavior. As an example, let's say you call Butch. He is slow to respond and so you greet him with a reprimand. But Butch doesn't connect the delay with your anger; to his mind, he came and you were upset. Maybe, he reasons, it is better not to come at all next time you call.

Sometimes, encouraging an alternative behavior can be a solution for some unacceptable ones. If your dog is begging at the dinner table, try putting him in a "Down-stay" and immediately reward him with praise if he obeys. (For obvious reasons, this is not a good time for a food reward.) In your dog's mind, you have rewarded him for his "Down-stay," a good

A choke or training collar, such as the one on this Great Dane, can cause physical injury to a dog if it is worn or used incorrectly. These collars can be a useful adjunct to training dogs and older puppies, but it depends on their size, strength, and temperament, and also on the skill and training of the owner. Ask a trainer if a choke chain is appropriate for your dog, and, if so, have the trainer teach you how to use it. If the human companion does not know how to properly use it, this type of training collar can cause serious windpipe injuries, especially to small breeds and shorthair dogs. In addition, a shy, sensitive, or submissive dog could be traumatized by the indiscriminate use of a choke chain. If your dog is a tenacious puller or has other problems with leash manners, consider instead using a head halter to correct his behavior (page 109).

reinforcement. Remember that what humans find unacceptable, such as begging, may be, from a canine perspective, perfectly normal and understandable. By avoiding punishing him for begging, which might have confused him, you turned the situation into an opportunity for obedience and praise. If, on the other hand, you had given your pet "just one" treat to get him to stop begging, you would have rewarded the bad behavior, and the dog would soon be asking for more. Every dog is unique, and what works for some may prove ineffective with others. If your dog continues begging after a few tries at eliciting an alternative behavior, you might want to resort to some mild punishment in the form of a "time-out"—removing the dog from the room for several minutes, and closing the door behind him.

For all commands, except for "Stay," say the dog's name first to get his attention and then give the command. Calling his name before "Stay" may be confusing to your dog, as his first reaction may be to go to you. Strive for simplicity, clarity, and consistency in commands. For instance, if you want your dog to quit investigating some questionable find, say, "No!" or, "Leave it!" Whichever term you choose, be sure you and your family use that same one all the time. Better yet, add a hand signal and give your dog two stimuli to perform the right behavior. Signals also mean you will be able to control your dog even when you cannot communicate by voice.

When you do give a command, be direct. State what you want clearly; don't bury the command in, what is to dogs, a mish-mash of other meaningless sounds. If you try something like, "Come on, Fido, how about showing us what a good sit you can do?" poor Fido is swimming in a sea of sounds, looking for one he recognizes—and you are becoming frustrated at his lack of obedience. Instead, say, "Fido, sit!" in a tone that expects, rather than requests, compliance and the dog will respond.

When you start training, food is usually the first and best reward, especially for puppies. Play is also effective, especially when a dog has just eaten, and is ideal if he is a bit overweight. For dogs who love a good game of tug, twenty or thirty seconds of it can be a potent reward. As training continues, reduce treat or play rewards and replace them with praise and affection. By also giving an occasional food reward, you can powerfully motivate a dog to obey, since he never knows when an extra treat may be forthcoming. Always end a training session on a positive note. If a new command has proved temporarily beyond your dog's current performance level, go back to a well-known command and reward compliance. Keep training sessions reasonably brief. A puppy's attention span is quite short, so measure early training periods in minutes. With an adult dog, be sensitive to signs of distraction or weariness. Stop and let the animal relax with a toy. Remember that all training sessions do not have to be formal. Reinforce learning by incorporating obedience commands into daily life. Asking a dog to respond to some command before being fed strengthens his obedience and reminds him who is providing all of the good things in his life. It's a good idea to vary the commands, otherwise the dog may automatically begin to "Sit" immediately before dinner. This way, he is forced to pay attention to you to learn what he has to do.

IN THE BEGINNING . . . SOCIALIZATION

Early socialization creates the basis for the "civilized dog." Experiences and learning during your dog's first six months affect him for his entire lifetime. Expose your puppy to as many new and varied experiences as possible—people, dogs, other animals, activities—while maintaining his comfort level. He should be excited, but not overwhelmed, by the new opportunities. As his world broadens and is filled with pleasurable events, the pup will become more and more confident and reliable.

Part of socialization is turning a nipping, wriggling, self-involved little pup into a family dog. Handle your dog gently but firmly so he becomes used to being groomed, cleaned, and checked. A little time spent now overcoming reluctance to have nails clipped or ears cleaned will pay dividends later to you, your vet, or the groomer. Soothe him by saying, "Easy," or, "It's okay," if he seems nervous. He will begin to connect the words with events that are not too scary after all, even if they challenge him at first.

Biting and bolting are two puppy behaviors that deserve early notice. Being nipped occasionally by razor-sharp puppy teeth is probably unavoidable. When your puppy starts to bite down, say, "Ouch!" and stop playing with him. If he doesn't get the message, call a brief "time-out," removing the puppy from the room for two or three minutes. He will soon connect the pressure of his jaws with the unpleasant experience of losing your attention. To meet his need to chew, provide him with appropriate soft toys. If he is having discomfort during teething, freeze a wet washcloth for a chewie that will also numb his gums.

Puppies also need to learn to inhibit the impulse to charge through doors or scamper away from you on a whim. At a doorway, gently restrain your pup by holding his collar or leash, and say, "Wait." As he learns to wait and to remain with you on command, praise him immediately.

When the puppy is eating, stay near him and occasionally remove his food dish, just for a few seconds. Add a tasty morsel to the dish and return it. The puppy will learn quickly that giving up his bowl to you brings rewards. Do the same with various toys, saying, "Out!" or, "Drop it!" When he releases the toy (as opposed to losing it through a tug of war with you) say, "Good dog!" and, after a moment, return the plaything, or give him a substitute if the original was not appropriate for puppy teeth. You are creating a connection between the word and the action so that later work with the "Out" command will go smoothly *(page 112)*. "Out" can be very important if your dog picks up a dangerous or valuable item you want him to

Teaching "Sit": The trainer holds the treat directly above the dog's head. As the dog's eyes follow the food, his head tilts back and his hind legs begin to fold under him to offset his body position, causing him to sit naturally. The trainer says, "Sit," as the dog's back legs begin to bend. The instant the dog is fully sitting he gets the treat. Once the sit command is understood, reduce and then eliminate the food reward, always praising the dog as he sits obediently.

release immediately. Also, start using the term "No!" (or whatever word you choose) when you remove him from unacceptable situations, and teach him the meaning of "Off" *(page 108)*. Although this may not seem like formal obedience training, the pup is learning the house rules and you are again reinforcing your position. If your dog begins threatening people who approach him when he has food, bones, or what he considers toys, consult a qualified behaviorist *(box, page 116)*.

Most adult dogs dance with anticipation whenever they see the collar and leash, but your young bundle of canine joy may rebel against them. The young puppy does not yet connect these items with a walk and, accustomed to being relatively free and unhindered, may be surprised and resentful at the intrusion of a collar and leash. Put his collar on comfortably and let him get used to having it around his neck, then attach a short leash and let him drag it around the house (watching that it doesn't snag on anything) until it, too, becomes normal. This can take from a few hours to a few days. When you do pick up the lead, jolly your pup along, encouraging him to come with you. Try to avoid a pulling contest. Instead, aim to show him that going your way will be a rewarding experience.

The best collar for all puppies and most adult dogs is a flat leather or nylon model. These should provide sufficient pressure for walks and training. You must be able to slide two fingers between the collar and neck (or measure his neck and add two inches for correct collar size). Make sure the collar is snug enough so that it can't pass over the dog's head, but loose enough to allow him to breath easily. Many trainers use the metal chain or choke collar, or even the prong-type collar which pinches the dog's skin when tightened. However, either of these collars may damage a dog, especially a puppy, if utilized incorrectly. Before opting for one of these training tools, seek the advice of a trainer on exactly how to fit and use them.

OF CRATES AND HOUSE-TRAINING
AND CONTROLLING DESTRUCTO-PUPS

In the wild, canids find their shelter and security in dens. Properly handled, crates can serve the same den function for domesticated dogs and also are a real aid in training. However, a crate is not really a den: In a den, no one closes the door. With some dogs, crates can be harmful. Dogs suffering from separation anxiety *(page 115)* can seriously injure themselves if crated during a severe panic attack. Their level of terror can lead to frenzied behavior and a lack of awareness of pain.

To prevent any negative responses to the crate, never use it as punishment and don't overdo crating. If a puppy needs to stay alone for more than three hours—or an adult dog for more than five hours—leave him instead in a doggy playpen or a blocked-off and puppy-proofed area of a

Opposite: Teaching "Down": While he's in the "Sit" position, hold a treat near the dog's nose and move the treat downward. As his nose follows the treat, move your hand forward in front of his face. Give the command, "Down," as he starts to lower himself, then move your hand further down and forward until the dog is fully lying down. Immediately praise him with, "Good, down!" and reward him with the treat.

Teaching "Stay": During early training for "Stay," the trainer should use a short leash so the dog is aware of the trainer's control. With the dog in the "Sit" position, the trainer says, "Stay," and, with palm facing the dog to reinforce the "Stay" command, moves a small step away while maintaining eye contact. After a brief pause, the dog is rewarded with praise and a food treat for remaining in position. Don't be too effusive in praise or say his name, as he will probably break his stay to come to you. Repeat the exercise, gradually increasing the length of the pause. If the dog breaks the stay, make him sit and repeat the exercise with a shorter pause before giving the treat. Once the dog masters this, the trainer uses a longer, slack leash to maintain control and gradually moves farther away as the dog remains in the "Stay" position. Eventually, as in the picture at left, the dog will maintain a stay solely by the verbal and/or hand command even when off-lead. To teach "Downstay," repeat this exercise, beginning with the dog in the "Down" position.

Releasing the dog from "Stay": First, with the dog in the "Stay" position, reward him for his stay with a treat. Then, standing in front of him, open your arms and say, "Okay!" or your own personal choice for a release word. You may then reward this behavior with a treat.

room with a small dish of water, some toys and a treat or two. A dog's crate should represent a warm, snug, peaceful nest for sleep and a safe, secure place to stay when alone. Let your dog adjust to the crate in stages, first in your presence, with the door open. The next step is to leave the room, and later, to close the door to the crate, gradually increasing the length of time you leave him alone. Give him special toys to be played with only in the crate, and feed him inside it at the beginning. (Don't leave food or water in the crate during the house-training process. Once your dog is trained, you can leave water for him, but buy a device that attaches to the crate to hold the bowl so it won't spill). A crate-trained dog has a safe place in which to ride when traveling, a quiet place to recuperate from illness or injury, and a sanctuary during hectic times around the house. Each time you put your pup or dog into his crate, say, "Crate," and soon he will obediently and quite happily enter his little "den" when you give the command.

A collar can be dangerous if it catches on the wire of the crate; always remove it before your dog goes inside. Put the crate in a room where you spend time, but instruct family members to leave the dog alone when he chooses to go in by himself. Moving the crate into your bedroom at night will help make your pup feel more secure.

In the wild, canids do not eliminate in their dens, and domestic dogs show the same aversion to soiling where they sleep. The crate allows you to use this innate inclination to help house-train your dog. Choose a crate that is just large enough to hold a dog at his adult size. If your pet is a puppy, you should block off the excess space with a piece of plywood or a cardboard box until he reaches his full size. That way he won't be tempted to use one corner as a toilet area. It's best not to line the crate with anything. As a general rule, a puppy can "hold" urine and feces for the same hours as his age in months, plus one. So a three-month old pup will usually stay dry and clean for four hours. Try not to crate him longer than that or he may become accustomed to living in his own excrement and be harder to house-train. This is a problem often encountered with pet store dogs and shelter dogs that don't get walked regularly. If you must leave your dog longer, do not put him in the crate, but use the puppy playpen described above, leaving some newspaper on which he can relieve himself.

Unless you plan to have your dog toilet indoors on paper (an option if you live on the twentieth floor of a high-rise apartment), adding a paper-

Teaching "Come": The dog should already know how to sit and stay. The trainer moves away, turns to face the dog, calling him and saying, "Come," while motioning with palm inward. In this picture, the dog has just begun to rise and move toward the trainer, while being verbally encouraged. Notice the trainer's hand position and the slack leash, which ensures control. When the dog reaches the trainer, he will be commanded to sit, and then will get his food reward.

Teaching "Heel":With the leashed dog in the "Sit" position on your left side, hold the treat in your right hand. Say the dog's name, followed by, "Heel," then start walking, left foot first, holding the treat just over the dog's nose and doling out little bits to encourage him. Keep the dog on a tight leash. Ideally, your dog should pace himself so that his shoulder is in line with your left knee, as shown above—reward him with larger treats when he does this.

training step to house-training usually just prolongs the process. If your housing circumstances require a paper-trained dog, get him to relieve himself only on paper in a specific place, and not on any piece of paper lying around the house.

Just because a puppy can hold his urine for several hours, does not mean that he will, especially if he is not in the confines of his crate. The secret to quick and relatively painless house-training is to catch the pup when he is just starting to think about going. He will need to go upon waking up, before entering his crate, after playing, within fifteen minutes of eating, and whenever you see him circling and sniffing. Take him outside immediately to the place you have pre-selected and that you plan to use consistently. Say the phrase you have chosen, "Pick a spot," for example, and praise him when he relieves himself. Especially for young puppies, be careful not to praise until he is finished, or he may become so excited that he forgets to do so. If he has an accident in the house, don't punish him. Keep an odor-neutralizing product on hand *(page 208)* and try to watch the pup more closely in the future. Adult dogs can be trained the same way but will not, of course, need to relieve themselves as often. Be patient and consistent.

Dogs don't know the value of material things people may hold dear, a fact that can spell disaster for your possessions. An energetic dog left alone for even a short time can wreak an amazing amount of damage. Here again the crate can be a salvation, keeping both the dog and your house safe when you cannot be there to supervise. Until the pup learns some limits, he should either be supervised or safely crated all the time. One handy way to maintain control is to tie his leash to your belt; this "umbilical cord" method works best if you are not moving around a great deal. If he starts to chew or destroy something, immediately say, "No!" When he stops, praise him and give him a toy to chew. If he climbs on the furniture, say, "Off," or whatever command you have chosen, while luring him down onto the floor with a food reward or an enticing toy. Put him in a long "Down-stay" before actually giving him the treat or toy. With timely input and reward, the puppy begins to understand which behaviors are acceptable. If jumping on furniture increases, the dog is likely getting reinforced for jumping on rather than off the couch. If this happens, set him up to commit the very same misbehavior while keeping him on a slack leash. When he attempts to jump up on the furniture, say, "No!" firmly and restrain him if necessary. Pause for a few seconds (or minutes, depending on the frequency with which he tries), ask for a "Down-stay," and praise him for not jumping. After a few tries, "No!" (without the restraint) should do it. If you catch him every time, your presence alone should eventually stop him from jumping on the furniture altogether.

Even after your dog has become trustworthy with your belongings, try to keep temptation out of his way, picking up small items he may view as toys.

OTHER COMMANDS

Once you and your puppy have mastered these commands through rewards for certain behaviors in day-to-day interaction, you're ready for some other essential basic commands, but they require a little more formal effort. (For these training sessions, keep your dog on-lead, and where possible, consistently on your left-hand side.) "Sit" forms the base of any training *(page 103)*. Obeying "Sit," your dog learns to control his behavior, and you have the chance to reward him for stopping a prohibited activity and sitting instead. This will come in handy when you want to keep him away from the kitchen counter during food preparation, away from the door when visitors arrive, or under control when meeting new people. "Down" puts the dog in a more relaxed position in which he can remain comfortably for a long period *(page 104)*. "Stay" can be used with "Sit" or "Down" to keep the dog in position *(page 105)*. Choose a word like "Okay" to release the dog from these commands *(page 106)*. "Come" should bring your dog rapidly to you *(page 107)*. "Heel" makes walking your dog a pleasure *(opposite)*. There are numerous other commands and even tricks you can teach your dog if you are patient and he has an aptitude. But with these basic commands, you will have an obedient dog that is a pleasure to have around.

A BETTER WAY TO LEAD

 Unlike conventional flat or choke collars, the head halter combines a nose loop and neck collar; the two loops come together under the dog's chin. The leash attaches underneath, where the two sections of the halter meet. The principle of the head halter is simple: leading a dog by controlling his head works better than trying to hold back a boisterous, unruly, or strong dog, especially for children and smaller (or older) adults. (By comparison, a choke collar provides a startling, painful yank to inhibit the undesired behavior.) However, although the head halter resembles a muzzle, it isn't one, and should not be relied upon to provide the same protection with aggressive dogs.

Since the nose loop in particular may feel strange, encourage canine cooperation with treats and praise. Fit is the most important factor in successful head halter use, so ask the salesperson or your trainer for help selecting the right size and making adjustments. If the dog seems upset and tries to remove the halter, distract him with play or a treat until he appears to be comfortable.

When using the halter, the bywords are "minimal force." In fact, yanking hard on a head halter can actually injure your dog. If he drags or lunges, pull the leash gently but firmly. As he, of necessity, quits the inappropriate behavior and moves into correct position, quickly release the pressure and immediately praise him.

Make sure the loop around the dog's neck is tight enough so that only two fingers fit under it, but keep the nose loop loose enough for the dog to be able to eat, drink, yawn or even bark.

PLAY &
·EXERCISE·

In a favorite game, a Labrador retriever jumps to catch a ball. Use tennis balls or other soft objects to avoid harming the dog's teeth. Vary the game by throwing some bouncing ground balls, and some higher pop-ups. Most dogs enjoy games of catch and, by training your dog to return the ball and drop it on command, you reinforce obedience training.

All work and no play makes Jack a dull boy, so the saying goes. Fido, too, becomes dull without the mental and physical stimulation provided by play and exercise. These twin activities give the inquisitive dog a natural outlet for energy, alleviate boredom, and reduce destructive behavior. Properly handled, play and exercise not only meet physical needs, strengthening canine bodies, but they also cement the human-dog bond. Make play a reinforcement of obedience training by periodically stopping the action and giving a command such as a sit-stay; once the dog has successfully complied, release him and play again. In this way, the continuation of play is the reward. If you're using a ball or toy, enforce the "out" command, and then give back the object as his reward for giving it up *(page 112)*.

How much and what kind of physical activity is best? Just letting your dog out into the backyard does not provide the kind of interactive exercise he needs. Most dogs will not start running around on their own. For some dogs, even a daily walk is insufficient. Age, weight, health, and breed all factor in to the equation. Certain breeds have very high exercise requirements—some up to three hours per day—and this is an important consideration when choosing your dog. It isn't fair to expect a dog bred for herding sheep or pulling a sled to be content with a quick walk morning and evening simply to do his business. And you will probably pay the price for this with an unhappy dog who may turn to destructive habits to alleviate his boredom. Do some research into the exercise needs of the various breeds *(pages 176 to 203)*, and take your lifestyle into account before taking on a very active dog. Once you choose your ideal breed, make a commitment to provide your pet with the proper daily workout.

In some cases, there can be too much of a good thing. Young, large-breed puppies face particular dangers from over-exercising, and may need frequent, but short, play sessions. During their first year, such breeds experience rapid bone growth and, if subjected to heavy or extended physical activity, can suffer orthopedic trauma and stress, disrupting their growth patterns. As always, your veterinarian knows what's best for your dog.

Walking, hiking, jogging, and running are generally excellent exercises for dogs (and their owners). Before starting a jogging or running program, have your vet check that your dog is in good shape and make sure he has been trained to heel and not to sniff every post *(page 108)*. Otherwise you'll experience constant trip-ups and unexpected stops. On the other hand, when walking, consider having a real "dog walk," letting your pooch investigate and enjoy all those smells and messages that are beyond our senses. This also provides a good opportunity for short, five-minute training sessions, to reinforce the basic commands, such as heel, sit, stay, down, and come. In warm weather or for any extended walk, be sure to bring water and check your dog's footpads for cuts, cracks, or tears, especially on hot or rough surfaces. Hot pavement can actually burn your dog's pads. If your travels take you into the woods or on the beach, check the area rules about leashes, etc., clean up after your dog, and avoid confrontations with wildlife or nesting birds for the sake of both your dog and the local inhabitants. For any high-impact activities, be sure not to push your dog beyond his capacity. Dogs generally don't monitor their activity levels and will continue, sometimes to their own detriment.

Swimming is an excellent exercise for dogs. Most take to the water if introduced as puppies, and some breeds—any of the retrievers and the Newfoundland—are naturals. Swimming is a no-stress exercise that increases flexibility and muscle strength. It is particularly well-suited for older, arthritic, or obese dogs. Even dogs suffering from paralysis are often able to regain partial use of their limbs in the water. Again, let your veteri-

Two curious Jack Russell puppies explore a rope toy. The soft material appeals to the teething youngsters' need to chew (and saves your socks from destruction). This could also turn into a puppy tug-of-war, strengthening muscles and improving reflexes.

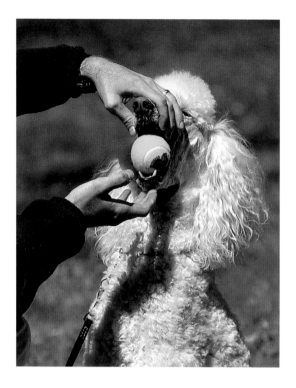

Teaching "Out": Approach the dog with a toy in one hand and a treat in the other. Encourage him to take the toy in his mouth. When he does, show him the treat, say, "Out," or, "Drop it," and take the toy. Immediately give him the treat and praise him, then return the toy, so that this doesn't signal the end of the play session. Some dogs willingly give up objects, but if not, you may need to resort to a hands-on technique, as shown above. Say, "Out," or, "Drop it," as you gently open his mouth to take the toy, then reward him with a treat and praise. Return the toy. (If the dog displays any threatening behavior whatsoever when you try to get him to release something, don't attempt the hands-on approach. Leave the dog alone and seek help from a professional behaviorist.) When you want to end the play session, repeat the "Out" command procedure, then allow the dog to wind down with calmer play. You can also use the "Out" command to exchange an inappropriate chew-object—such as your shoe—with an appropriate item—such as a rope toy.

narian be your guide. Older or injured dogs as well a young pups can don a canine life vest to assure a safe swim, while allowing for the beneficial exercise. If you do swim with your dog, be sure to rinse him off well, especially if he was in salt or chlorinated water, and keep him warm and out of drafts until he is dry.

THE GAME DOG

Dogs love games; not surprisingly, the games they enjoy most are those that play into their own natural drives and instincts. Catch and fetch are winners. Be sure to use a soft ball, such as a tennis ball, to avoid harming teeth and vary the fun by using a frisbee (obtain a special dog-designed one that will not hurt canine mouths). Be careful with sticks. Although popular with dogs, they can become a choking hazard if bits are chewed off, or dangerous if the stick splinters. When the idea is to fetch rather than catch, elderly or obese dogs can join in. And, even with younger dogs, be careful to throw low to the ground. Those spectacular up-in-the-air twisting catches are great to watch but can sideline your dog with muscle strains and even spinal problems. You can try playing hide-and-seek, or soccer with a large ball. Another great source of fun and safe exercise is to join a local dog agility club. Contact your veterinarian, dog trainers, or animal shelters to find out where an agility club meets in your area.

Other games, such as tug-of-war, wrestling, and chase—games that dogs seem to like a great deal—are controversial. In the past, some experts advised against playing them with dogs, believing that excitement could escalate into aggression. Currently, however, most behaviorists do not believe tug-of-war causes aggression and they see no harm in a little rough play once in a while, as long as both parties enjoy it and dog owners use common sense. Of course, dogs with a tendency to nip and bite should not be encouraged to wrestle with humans. But plain old rough play, a study shows, is in general simply a temperament trait: Some dogs get into it more than others and some can get overexcited. Any injuries that result from roughhousing, especially when a large dog tussles with a small person, are almost always accidental.

In any game you need to be sensitive to signs that your canine playmate is taking things too seriously: nipping or biting, snatching or guarding toys, refusing to let go of objects, a raised lip, or serious growling are all warnings. (Many dogs growl in the heat of play, but listen for a change in tone from playful to threatening.) If any of these signs occur, stop the game immediately and reconsider playing that particular activity with your dog. If it happens again, consult a certified behaviorist (*page 116*).

Most dogs enjoy playing with other dogs, especially if they did so when young. Females tend to play readily; unneutered males are more likely to get into aggressive encounters. Introduce potential playmates on neutral territory, and watch for signs of hostility. Even obedience-trained dogs can get into fights, and once an altercation begins, even your loudest and firmest "No!" will usually be ignored. Try to find compatible playmates for your dog and set up future playdates with the most successful matches.

There are also times when your dog plays alone, which makes your choice of his toys particularly important. Avoid small toys he could swallow. Even large toys may prove dangerous if they have have strings, elastics, or ribbons that can be pulled loose or squeakers or other small parts, such as eyes, that are eagerly and easily removed by sharp teeth. Look for larger, sturdy toys that won't break into pieces. It is also a good idea to avoid using old shoes, socks or gloves as toys, since your pet may find it hard to differentiate them from your good ones, creating a real communication problem between you. The best toys are safe, fun, and designed to help your dog release energy and amuse himself. Good choices include nylon bones, hard rubber toys, rope toys, and sterilized hollow bones. Many dogs also enjoy rawhide toys, although occasionally a young pup will overindulge and suffer an intestinal blockage. Supervise your dog's rawhide intake to see if he has any problem tolerating it, and discard the toy when it becomes small enough to swallow.

DOGS UNLEASHED

Dog parks allow pooches the chance to play off-lead with fellow canines. But, before you let the games begin, keep the following in mind, then get ready for some wild and crazy dog antics!

◆ Does your dog need an obedience refresher? Don't wait to find out at the park that his off-lead skills aren't up to snuff.

◆ Is he too aggressive or shy for this experience?

◆ Know park rules. Even if not required, make sure that your dog's vaccinations are up-to-date.

◆ Check the fencing. Unfenced or inadequately fenced parks expose dogs to traffic and roaming animals.

◆ Clean up after your dog. And, refill your digger's holes.

◆ Have him wear a flat collar with clear identification.

◆ Bring water and a bowl, a leash, toys you're prepared to lose, towels—and a camera.

A secure fence around this field ensures that these off-leash dogs are safe to romp and run without escaping into traffic or other danger.

BEHAVIORAL
·PROBLEMS·

Even with obedience training, dogs do not always act as you would wish. You may find yourself feeling angry, disappointed, and frustrated, wondering how your dog could do this to you. Unfortunately, such thinking often backfires, and hard as it may seem, you need first to defuse any emotional reaction to the dog's "misbehavior." When you approach behavioral issues calmly, without trying to blame or punish, you will have gone a long way toward a solution. The next step is to decide to be active (like a pack leader), rather than reactive. Define exactly what is happening: What is the real problem? When does it occur? Are you doing something unintentionally to reinforce or encourage the unwanted behavior?

You'll want to maintain this mind-set if you have to deal with any of the "dreaded dog deeds" that follow, and you'll also need to recall some basic dog psychology. First, dogs, living largely in the present, connect rewards and corrections with their most recent behavior. Thus it is easy to contribute to a dog's misbehavior by inadvertently rewarding the behavior—some dogs are reinforced by attention alone. By talking to them, gazing at them or even through our posture, we can send the wrong message. We can also deter a dog's obedience by unintentionally discouraging it.

Second, when dogs "misbehave," remember that they do not perceive their behavior the way you do. Dogs don't think in terms of "good" and "bad"—

The natural curiosity of a puppy left unattended and uncrated can result in disaster. To the dog, your precious antiques have the same appeal as his chew toys and a houseplant yields a very satisfying crash as it falls, releasing some interesting dirt that can be spread around for even more amusement. If you catch the offender in the act, correct him with a firm, "No!" *(page 116)*. If you come upon this scene of destruction after the fact, don't discipline the pup; he won't connect the past half hour's play with your present anger. Instead, resolve to keep him in a puppy-proof crate or pen when he is alone and provide him with appropriate toys and adequate exercise.

there is no intent behind their so-called misbehavior. Playing by the dog rules doesn't include respecting material possessions or landscaping, nor is howling or barking considered unacceptable. This is not to say that you cannot teach your dog to live obediently with you; it just means that you can't expect your dog intuitively to know what is okay and what is not.

Third, as social animals, dogs were not meant—and do not like—to spend most of their lives alone, while you are away at work or play. A dog owner must respect the dog's need for interaction and make arrangements to meet those needs, either by being there or by providing other canine or human companionship, such as doggy day care, a midday dog walker or, if appropriate, by getting a second dog *(page 75)*.

Finally, while dogs and people have lived together for millennia, they haven't always understood each other, in spite of the best efforts of both species. In a lot of cases, you can get your dog back on track by refreshing his basic training *(page 100)* and upping his daily exercise; in general, a tired dog is a good dog. Crating *(page 105)* can be a solution, albeit temporary, to many problems, although it can be harmful to some dogs suffering from separation anxiety. If you feel at an impasse, turn to the professionals. Consult a certified applied animal behaviorist *(page 116)* or hire a dog trainer for some private sessions. Sometimes a little timely third-party intervention can turn a potentially bad situation around fairly quickly. Seeking help is not an admission of failure, but an investment in the future for you and your dog.

ANXIETY-RELATED PROBLEMS

Dogs, like humans, can develop irrational fears or phobias. Sometimes it is clear where the problem began. A dog threatened or frightened by a man with a beard may show fear and either attempt to escape or growl when encountering other men with beards. Dogs seem particularly susceptible to noise-related fears. Many dogs show severe anxiety (shaking, salivating, cowering, whimpering) during thunderstorms or when firecrackers are exploding nearby. Dogs also seem particularly vulnerable to developing anxiety when separated from their owners. This can be a serious problem, and is often misunderstood and misdiagnosed. Dogs who have separation anxiety become extremely distressed when family members leave them alone. The anxiety they show is similar to a panic attack in humans. Some injure themselves in frantic attempts to get out. Dogs in this state have jumped through plate glass windows, dug through doors with their claws or broken their teeth and ripped their paws trying to get out of a crate.

Separation anxiety is expressed differently in different dogs. Some dogs bark continuously when their owners are away, others become destructive and do considerable damage with their desperate scratching and chewing. Still others appear to lose their house training and eliminate in the house. Owners often assume their dog is angry at being left alone and believe the

If left for several hours, dogs enjoy the additional space provided by a canine playpen (or play area blocked off by baby gates) more than a crate. The puppy can move, stretch, and play happily with his toys without putting himself or your possessions in danger. Just remember that as puppies grow, today's barrier may tomorrow be merely a minor obstacle to be jumped or scaled.

behavior during their absence is an expression of this anger. Confronting the damage, they often punish the dog. This doesn't help an anxious dog. The frenzied behavior that occurs is not really under voluntary control and the problem can only be solved by treating the anxiety.

Recent research has explored the use of human antianxiety medications for use with canine anxiety problems. These have been shown to have some effectiveness, but usually the best results are achieved with behavioral interventions supplemented by medication when the problem is severe. If a dog who has been house-trained starts eliminating in the house, or if he starts scratching or chewing and destroying things, and this occurs only when family members are away, there is a good chance that the behavior is related to anxiety. Sometimes, when the anxiety is moderate, if a dog has been accustomed to a crate and is comfortable in it, crating can have a calming influence. It is tempting to use a crate if the dog causes considerable damage during owner absences, but this should be approached cautiously. A crate can limit the damage to the household but is unlikely to eliminate the anxiety. And when anxiety is more severe, there is a risk that the dog could seriously injure himself if left in a crate.

AGGRESSION

Canine aggression is a serious problem. Nobody likes to be physically threatened by a four-legged family member with the teeth and jaws to back up those growls and snarls. Meeting aggression with aggression can escalate a confrontation into a tragedy. At the first sign of inappropriate behavior, seek help from your vet to eliminate the possibility of physical causes, and consult an animal behaviorist or a trainer to assess the situation and develop a plan to modify human-dog interaction. There are different kinds of aggression, and the treatment for them differs depending on the kind. If the problem is serious, skip the trainer and consult a qualified behaviorist right away. Of course, the best way to treat aggression is prevention: unless you are an experienced dog owner, don't choose an aggressive breed and avoid play biting. But most important of all, socialize your dog well *(page 103)*. Don't allow your puppy to growl and snap because it's cute; discourage that from the start.

DESTRUCTIVE BEHAVIOR

A bored, anxious dog seeks an outlet for his energy and a release for his stress. When combined with a lack of exercise, this is a prescription for an unwanted remodeling of your house. By the time you discover the chaos, your dog will not connect any corrections you make with the behavior that disturbed you. If your dog does any damage in your presence, correct him and distract him into acceptable activities. Until he understands his boundaries, keep him either supervised or connected to you by leash *(page 108)*. If the damage occurs in your absence, crate your dog (unless the problem may be due to separation anxiety) or restrict him to a truly dog-proofed area. Try leaving a radio on, hiding toys for him to find, or hiring a dog-walking service to break up his day. If destructive behavior continues, consult an animal behaviorist or trainer for advice.

Since a dog's teeth are his primary means of wreaking havoc on your possessions, chewing is a subset of destructive behavior. Always provide your dog with stimulating and appropriate chew toys *(page 113)*. During the discomfort of teething, which can extend up to two years, your dog has a special need for chewies for relief.

DIGGING

Some breeds, such as terriers, are especially energetic diggers by nature. Correct your dog if he starts digging in inappropriate areas, but since digging is an instinctive activity, provide him with a fenced-in, free-dig zone of his own, and teach him the "Dig" command. In your absence, limit his access to other areas of the yard. Provide other outlets for his physical energy *(page 110)*.

BARKING

All dogs bark, but as with digging, some breeds bark, howl, and generally vocalize more than others. Keep this in mind when choosing your pet. If your dog barks excessively, try immediate correction and diversion to another activity. However, if your dog perceives a human's verbal reprimand as welcome attention or even a "bark-a-long," he may decide to bark more. Try teaching a "Speak" command. Once this is learned, teach "Quiet:" Divert your dog's attention by taking him to another room, and after a few seconds of quiet, reward him for his silence. Unfortunately, a lot of problem barking occurs when you aren't there to hear it, but your neighbors are. Separation anxiety is often the cause of such barking. If your dog is outside during the day, block his view of the neighbors' properties so that visual cues will not start him barking. (Of course, sounds from a neighboring yard may still set him off.) You could try a bark-activated collar that delivers a whiff of citronella (generally unpleasant to dogs) when the dog barks. If barking becomes a real neighborhood nuisance, and even professionals can't help, you may have to choose between giving up your dog or having him de-barked, a surgical alteration of the vocal cords to reduce the volume of sound. De-barking is controversial because many people consider it inhumane. It should be considered only when the other alternative is euthanasia.

FAILURE OF HOUSE-TRAINING

Even if your adult dog was never house-trained, it's not too late to start *(page 105)*. House-training an adult dog takes longer, but, with patience and consistency, it can be done. However, if your dog was previously house-trained and

There is nothing like a good dig, especially for breeds like terriers. This dog may be digging out of boredom or just for the fun of it. Since you probably do not have as much fun filling the holes in your lawn or garden, find your digger other physical outlets for his energy or, if he really enjoys digging, set aside a separate part of the yard where he is allowed to dig, away from expensive landscaping or delicate flower beds.

This unruly dog jumps up on his owner and generally does as he pleases. Try to avoid having to constantly use the negative commands "Off!" or "No!" Instead, when the dog jumps up, move back and say, "Sit." When he complies with your command, crouch down and reward his quiet behavior by calmly stroking his sides. Don't neglect the reward and praise once he is sitting consistently at your arrival or you may inadvertently encourage the dog to return to jumping for the increased attention. The idea is to get him to realize that by sitting he will get attention. If your dog ignores your commands, keep him on the leash at all times outdoors and revisit the basics of sit and heel until he is under control. Letting him run loose only exacerbates the problem.

starts to urinate or soil inside, visit the vet; incontinence is a symptom of many diseases and conditions. If the vet pronounces him healthy, consider remedial training, especially if the dog is still a pup. Return to careful supervision, escorting him to the same spot, and enthusiastic praise on compliance. Only crate him when he's alone if you can absolutely rule out separation anxiety as the cause of the house-training breakdown. If there are still mistakes, keep a careful log of when problems occur. Don't become angry or abusive. And don't get rid of the dog without taking him to a behaviorist or trainer who can advise on such problems.

JUMPING UP

Dogs often jump up on people simply as an enthusiastic greeting. Avoid talking excitedly or waving your arms when entering a room; this just encourages the dog's behavior. Instead, immediately command the dog to sit, and then reward this behavior (left).

SEX-RELATED MISCONDUCT

Urine marking, when a dog (usually a male) leaves a small amount of urine around the house, differs from house-training problems. It establishes the dog's territory and may occur even after he seems to have completely relieved himself outdoors. Neutering a male almost always eliminates all inside marking, and diminishes the frequency of outside marking too. If a dog is neutered early enough (page 160), marking can usually be prevented altogether. Spaying a female, on the other hand, has little effect. To quash this annoying behavior keep him on a leash in the house so you can correct any initial moves toward marking. Always thoroughly clean and deodorize any marked areas (page 208).

Dogs may also attempt to mount humans, although it is not clear why they do this. Keep the dog on-lead for control, say, "No!" when the behavior occurs and immediately isolate the dog for a few minutes. Avoid touching him as this rewards him and reinforces the behavior. Then command him to sit, reward compliance, and play with him. Followed consistently, this routine should reduce or extinguish the behavior.

EATING FECES

The dog's habit of eating his own or other animals' stools, called coprophagia, is particularly disgusting to people. Your dog, however, sees no problem with eating excrement, and herbivore feces may even provide nutrients. However, various internal parasites also may be ingested in this way. While numerous theories have been advanced to explain coprophagia, treatment is the same. First, confirm with your vet that your dog's diet is adequate; also ask for an additive for your dog's food to make his stools distasteful. Clean up immediately after your dog defecates and place cat litter boxes out of reach to remove temptation. If your dog tries to eat feces while on a walk, correct with, "No," or, "Out" *(page 112)*.

GARBAGE SCAVENGING

Again, while this practice offends human sensibilities, a dog's perspective is different. He is exploring and enjoying new tastes and smells. However, beyond the mess that the dog can make, garbage may contain dangerous chemicals, bacteria-laden food, and sharp objects that can cause serious harm. Deterrence works best; securely latch garbage cans and remove them from the dog's area. If you catch your dog in the act, say, "No," and reward his compliance with praise—not food. A head halter *(page 109)* can be a great help when walking your dog in areas where litter or refuse is present. With good reflexes, you can usually steer the dog's head away before he picks up a piece of garbage. If your dog is faster than you, say, "No," or, "Out" *(page 112)*.

BEGGING FOR FOOD

Your dog will naturally be interested in human food. It tastes and smells good, and besides, his humans are eating it. But that cute, begging puppy very quickly becomes a nuisance as his size and appetite increase. The only way to stop begging is to *never* reward your dog with that one little snack because he looks so adorable. Occasionally giving your dog a taste from your plate provides strong reinforcement for a behavior you want to stamp out. When your dog begs, put him in a down- or sit-stay some distance from the table *(right)*.

ESCAPING

To some dogs, a fence is just a brief obstacle on the way to adventures in the neighborhood. Sterilizing dogs will make them less interested in roaming. Since an open gate is the usual escape route, be sure everybody knows that it must be shut securely. If your dog is tempted by visual cues, try to block his view with landscaping or solid fencing. Making sure your dog is well exercised and has amusing toys will also help keep him in the yard. If you have a persistent climber or digger, consider stringing electric pet fencing along the top or bottom of your fence to deter him.

CHASING

Dogs like to chase fast-moving objects such as cars, bicycles, or other animals, and often get killed doing so. Confine your dog in such a way that he cannot chase cars. When on a walk, get your dog to heel or place him in a sit-stay to prevent chasing behavior, and praise him enthusiastically for his obedience. If he ignores your commands and takes off after a car, an in-line skater, or a neighbor's cat, intervene immediately with, "No!" and a tug on the leash sharp enough to make him fall back. Next, work on some remedial basic training to prevent this dangerous behavior from recurring *(page 100)*.

This well-trained dog holds a down-stay, allowing his owner to enjoy her dinner in peace. Obedience training means that the owner can praise her dog for compliance (or give him a reward, but with something other than food), rather than spending an increasingly tense and unpleasant meal continually telling her dog, "No!" Sometimes, however, even a usually obedient dog cannot resist the temptation of asking for food. In that case, it is better to crate the dog at the start of the meal.

KEEPING THE

⋄ PEACE ⋄

It should come as no surprise that, being the social animals they are, dogs enjoy company. A second dog, or even another animal, can become a canine playmate and antidote to boredom during times when an owner is away. But while such a buddy may help prevent destructive behavior, don't plan on a new family member curing obedience or behavior problems: You may just find yourself in double trouble. Instead, refresh your current dog's obedience work before you add a new canine and be sure any new member of the household is at least well socialized and, if possible, already obedience-trained as well. The likelihood of conflicts will be further minimized if you have your dog sterilized and choose a spayed or neutered dog of the opposite sex as the new pet. Dogs also generally adjust better if their housemate is a different breed, age, and size.

INTRODUCTIONS, CANINE STYLE

In the ideal situation, both dogs are well socialized to other dogs. Choose a neutral, fenced-in meeting ground like a friend's backyard to introduce your dogs to each other. Make sure there are no other dogs around. For now, keep both dogs on their leashes, even if one of the two has not been socialized. Have a friend hold the new dog's leash. Although some experts believe that being off-lead in this situation can prevent a fight from breaking out (by allowing one of the dogs to run off), it's probably not a good idea. If things get out of hand, you'll be better able to control the dogs and the entire situation if you and the person helping you are holding the leashes. If either dog does behave aggressively, turn his head and try to distract him with a toy. If the initial meeting goes smoothly, have the two meet on home territory, separated by a fence or a slightly open door, which permits them to scent each other but not come in physical contact. If they seem to be progressing, allow them off-lead in your presence; they will probably proceed through some ritualized aggressive behaviors, establishing dominance. Try not to be anxious about their behavior, because dogs easily read human body language and may respond to unintended emotional signals. Their "play" may be noisy and boisterous to the human eye and ear, but you shouldn't intervene unless truly serious fighting and biting begin. If you do have to separate the two, remember that an agitated dog, whether aggressor or victim, may bite you in the heat of the moment. Have some human help on hand, as well as a hose or water pistol to aid in breaking off attacks.

Meeting in neutral territory, a Samoyed and a Staffordshire bull terrier engage in some initial testing, each staring a challenge at the other. Keeping both on a leash for this first meeting prevents a confrontation from escalating to a fight. The desirable outcome is that one dog will avert his eyes and acknowledge the other's status, setting the stage for a play relationship.

Opposite: Two coyotes challenge each other. One may be an intruder on the other's territory, or they may be youngsters checking out their strength with a mock confrontation. With raised hackles, intent stares, and open mouths displaying bared teeth, they are exhibiting some preliminary aggressive posturing. Domestic dogs may adopt this same stance, and unless one backs down, a fight will most likely ensue.

Once the dogs are separated, grab their leashes and walk them away from each other. If the dogs seem to just be play-fighting, let them work things out. Human intervention is unlikely to alter the balance of power anyhow, so resist the natural temptation to favor the underdog. As long as the aggression is not serious (no full-fledged fighting or injuries), the dogs will eventually come to an understanding.

To limit disagreements, be sure each dog has his own food bowl. Feed them together, if possible, but with the bowls placed so the dogs are not facing each other, setting the bowls far apart if necessary. Most dogs will share a water bowl, however. Each should also have his own sleeping area and toys, but be prepared for each wanting the other's toy, even if they are exactly the same. If aggressive behavior continues, try to identify the precipitating events or circumstances and avoid these situations. Ultimately you may determine that the animals have to be kept separate when you are not home to keep the peace. Your veterinarian will be able to counsel you and may suggest a visit to a behavior professional *(page 116)* to learn the best strategies for reducing intraspecies aggression.

Introducing a young pup as the second dog in the house is somewhat easier. Try not to let the puppy's arrival upset the other dog's routines. Even so, the older dog's behavior may temporarily regress and, until the initial adjustment period is over, you may feel like you have two puppies, one large and one small. If you gently but firmly enforce the older dog's obedience, he should soon return to normal. However, unless truly threatening, do not correct the adult's aggressive growls toward the puppy; these reflect the normal adult canine's showing the puppy his place in the family. Since the puppy will be fed frequently, divide the adult dog's food intake into smaller meals so he will eat as often as the newcomer. Allow the older dog to sometimes escape from the puppy's attentions and enjoy some special time with you. Also be sure that he and the puppy have separate sleeping places, and do not allow the puppy to invade his elder's territory. Even if they become great friends, work with them individually when obedience-training the puppy or reinforcing the adult's training. Finally, remember that canine hierarchy is not static. As the puppy grows and becomes sexually mature, he may reverse roles with the older dog.

OTHER ANIMAL COMPANIONS

Dogs will admit other animals as members of their group—but the fact is, dogs and rabbits or rodents really weren't made to get along. Unless your dog was raised with these creatures from puppyhood, the canine predatory instincts may not remain permanently suppressed in the presence of small,

It may look like a serious fight, but this Labrador mix and German shepherd are merely involved in some healthy play-fighting, although some puncture wounds may result. You can recognize a friendly tussle by the sounds (the noisier the better) and by the typical pattern of biting around the head and shoulder area.

defenseless, running creatures. Cats are another story. With a little effort on your part, your cat and dog can live in mutual respect (or at least tolerance) for each other and may even become good friends. Keep in mind, however, that the two species do not always understand each other's body language, so some miscommunications must be expected. In general, it will work out better if the newcomer to such a situation is a kitten or puppy, or an adult who has formerly lived amicably with the other species. If you already have a dog, make sure he is obedience-trained before bringing in a kitten or cat, and have him meet some friends' cats to make sure his instinct isn't to attack all things feline.

INTRODUCING DOGS AND CATS

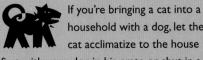 If you're bringing a cat into a household with a dog, let the cat acclimatize to the house first with your dog in his crate or shut in a room. For a home where a cat already rules the roost, be sure to give him lots of attention away from the new dog.

When you feel the animals are ready to meet, clip the cat's claws to protect your dog's delicate nose. With your dog leashed so you can pull him back if necessary, let the dog and cat meet. There may be no great

reaction on either part, but the more typical scenario is the cat hissing, arching his back with all his fur fluffed out, possibly slashing at the dog with his claws, but more likely beating a hasty retreat to higher ground.

Continue this routine, increasing the length of each meeting until both animals seem comfortable. Before you unleash an adult dog, make sure he will obey your verbal commands to stay away from the kitten or cat, if it looks like he might do harm. (A pup rarely poses the same threat, except

to a small kitten.) Do not leave the dog and cat alone together without supervision until you are confident that it is safe to do so.

Provide your cat with places he can hide, away from the dog. Cats like to get up high to survey their domain, safely out of a dog's reach. Make sure the cat can eat, drink, and use his litter box undisturbed by the dog. Place his necessities up higher than the dog can reach, or set up a "cat room," wedging the door open just enough to let your cat in, or cutting a cat-sized hole in the door.

Portrait of domestic peace: a dog and his cats or perhaps cats and their dog. Either way, these animals seem very comfortable together. One kitty reaches out a playful paw and the dog, knowing he is in no danger from sharp claws, leaves his sensitive nose unprotected. The cats lounge, perfectly relaxed, feeling no threat from their far larger housemate.

THE TRAVELING
· DOG ·

This happy poodle doesn't understand the potential danger he is in. Riding with his head out the window leaves this dog's sensitive nose, ears and eyes unprotected from flying objects. Dogs should ride in a crate, attached with a harness or, alternatively, lying down on the back seat. Otherwise, they might interfere with the driver and, in the event of an accident or even a quick stop, risk being injured or even thrown from the car.

There are likely very few words that your dog likes to hear more than "Car." Unlike cats, dogs seem to enjoy any length of journey, although pets that are ill, extremely shy and nervous, or very elderly generally respond better to alternative arrangements, such as pet sitters or kennels. But even healthy, mellow, well-behaved dogs may need some fine tuning to become road ready. Give a nervous canine traveler a snack as a reward for getting into the car; and, as preparation, move gradually from sitting in the parked car to short and then longer trips. Protect your car seats with covers or blankets in case of any car sickness, which should be dealt with calmly and casually. You can reduce the likelihood of tummy problems by feeding your dog a light meal two or three hours before leaving. Save his main meal for the end of the day. Of course, you should make sure he relieves himself before the trip starts and plan plenty of pit stops along the way, each including a drink and leash walk. This is always more safely accomplished at a rest area than along the side of a highway; be sure to attach the leash before opening your car door. And never leave your dog unattended in the car, especially when there is any possibility of heatstroke *(box, below)*.

Your dog will travel most comfortably and safely in a crate secured in your car. He cannot interfere with your driving, has his own space with a familiar blanket and toys, and will not be bounced around in case of an accident. Alternatively, consider a dog car harness, or for dogs under ten pounds, dog car seats; each securely attaches to a car's seat belts. If your vehicle has a passenger-side air bag, then always keep the dog in the back seat to avoid serious injury should the air bag deploy. A third choice, a divider insert, keeps your dog separate but does not anchor him safely. These are particularly handy for station wagons or other vehicles with a cargo area in back. In any event, dogs should never ride loose in a car with their heads hanging out of the windows. Flying debris can injure eyes, ears, and noses, and your dog is also at risk of being thrown from the car, or jumping out of the window to pursue a passing squirrel.

PREPARATION

Before departure on a long trip, you should confirm that your intended accommodations permit pets and determine whether any special rules or fees apply. Ten days beforehand, visit your veterinarian for a canine physical, including scheduled inoculations, and to obtain state-required rabies and health certificates. If you are planning international travel with your pet, contact the appropriate consulate for necessary documentation. When leaving the United States for a destination in Canada or Mexico, you'll only need rabies and health certificates. Some countries (although not Canada or Mexico) and even some states (Hawaii, for example) may require quarantine periods for visiting dogs which could cause you to rethink your plans.

In addition to travelling papers, your dog needs food, bowls, blankets, leashes, a pooper scooper and plastic bags, and any medication he may be taking. To avoid the unpleasantness of travelling with a dog with diarrhea, you can bring some of your own water from home and make the switch over to local water a gradual

Dogs who haven't experienced car rides as young pups, who become carsick, or who only go in a car to visit the vet may show reluctance to enter a car. This anxious dog is nervous about getting into the car, so his owner gently entices him with a treat. (If the dog suffers motion sickness, use a toy instead.) Avoid overexciting the dog, as this can lead to nausea.

HEATSTROKE

Heatstroke occurs when body temperature suddenly rises to dangerous levels. It poses a particular danger to dogs, who rely on panting to keep cool. (Only the canine's paw pads have sweat glands.) Overweight, elderly, and ill dogs are most at risk. Parked cars and other confined spaces with little ventilation represent danger. In humid, 75°F (24°C) weather, a car, parked in the shade with partly open windows, reaches 120°F (49°C) in just a half-hour and the dog inside suffers.

As his temperature rises, he becomes weak and uncoordinated; shock, coma, and then death may follow. If your dog is a victim of heatstroke, soak him in cool (not cold) water and rush him to a vet.

Heatstroke can be avoided by keeping your hot-weather dog in a cool, ventilated space with plenty of water. In really hot weather, avoid car trips if possible, unless the vehicle is equipped with air conditioning. Sun blinds attached to the car windows can also help keep your dog cool.

This eager passenger sits securely in a dog harness, which fastens to the car's seat belt. He is able to enjoy the view out the window without interfering with the driver or being a danger to himself or others in case of an accident. Riding in the front seat poses a potential danger to your dog if you have a passenger-side air bag in your vehicle. In that case, he should be harnessed in the back.

one. Don't neglect your dog's grooming while you're on the road; bring along any necessary grooming supplies. Your dog's favorite toys and blanket or pillow will be familiar and comforting to him. Make sure your dog is wearing a flat collar (and never a choke collar) with identification tags listing your home address, a neighbor's phone number, and a number at your destination. In the event that he gets lost, bring along a good photo *(page 209)*. And it never hurts to have a canine first-aid kit with you *(page 205)*.

DOGS IN FLIGHT

Dogs have joined the flying public, but—with the exception of very small dogs whose carriers fit under the seat in front—most dogs do their traveling in the cargo hold. While the USDA generally regulates live-animal transportation, you should confirm with your airline that your dog will be in a heated and pressure-controlled hold. Avoid flying puppies younger than eight weeks and pregnant, ill, or elderly dogs.

A crate is imperative for flying. Your dog should travel in a strong crate large enough for him to stand up in and turn around, lined with paper and clearly marked "Live Animal—This Side Up". Both the dog (via his flat collar and tags) and the crate should be clearly labeled with name and address. The crate should be unlocked and include, in case of delay, enough food and your dog's regular medication (including instructions) for twenty-four hours.

Try to fly on nonstop, direct flights. To minimize delays, travel at off-peak times. Feed your dog one-third of his usual meal four hours before the flight, and water and exercise him just before departure. Unless advised by your veterinarian, do not sedate him. Have his rabies and health certificates ready at check-in. There are numerous horror stories about improper air pressure or temperature in the cargo area of airliners. To be safe, well before take-off, you may want to talk to the boarding agent or a crew member so that the responsible individuals on the flight know a family member is down there.

If all of this seems too complicated or you think it will upset your dog, opt for a pet sitter or kenneling. A pet sitter makes your house look lived in and allows your pet to stay in his home environment. Although kennels take your dog away from his familiar surroundings, a good kennel can be a relatively low-stress environment *(box, opposite)*. If you choose either of these options for your stay-put pet, call the sitter or kennel as often as your peace-of-mind requires to check on the dog's well-being, and to answer any questions the sitter or kennel staff may have.

MOVING

Moving has its own special stresses for dogs, so your pet needs extra attention and reassurance during this time. Keep your dog in a safe spot, out of the action, on moving day. Once in the new home, a quick return to familiar routines and the swift reappearance of well-known toys and bedding should reassure your dog that he is home.

KENNELING

 You will want to select a boarding kennel with care. After all, an important member of your family will be staying there. Ask a fellow dog owner or your veterinarian for leads; then set up a visit. Inspect the premises for cleanliness and general upkeep. Are the runs properly designed, with sturdy fencing, solid dividers, good ventilation, screening against insects, bedding, and temperature control? Are there adequate sleeping quarters and space for exercise? Do the current tenants seem healthy and free of stress? Talk "dog" with the staff. They should show interest in and concern for their charges. They should require proof of your dog's vaccinations and ask about his temperament and feeding schedule. If your dog has a special disposition, is a little shy, perhaps, or tends to be a bit aggressive with other dogs, be sure to discuss this with the staff. You may find them very knowledgeable about handling dogs with attitude, but if not, continue your search for the ideal canine lodgings for your pet.

At drop-off, leave written instructions for your pet's care: medications, idiosyncrasies, and details of feeding. Be sure to provide sufficient food if your dog eats a special diet. Take along some of his favorite toys. Make sure the kennel has instructions on how to reach you, plus the phone numbers of a responsible neighbor or friend and your veterinarian. (Make prior arrangements with your vet to care for your dog if necessary during the period you are away.)

By carefully choosing a kennel, you provide your dog with good care while you are separated and guarantee a happy homecoming.

Individual outdoor runs at this kennel allow each dog to get fresh air and exercise without the risk of scraps with the other boarders.

CHAPTER

·5·

DOG
CARE

◆ ◆ ◆

"If you take a dog which is starving
and feed him and make him
prosperous, that dog will not bite
you. This is the primary difference
between a dog and a man."

◆

MARK TWAIN

FEEDING &
·NUTRITION·

"A dog starv'd at his master's gate
Predicts the ruin of the State..."

WILLIAM BLAKE

"If your dog is fat,
you aren't getting enough exercise."

UNKNOWN

How tempting it is to toss your leftovers into your dog's feeding bowl. Omnivores to the core, dogs will eat almost anything you give them (though they'll choose meat over veggies any day). But some forms of nutrition are better for them than others. Your dog may seem to be the ideal leftover disposal system, but don't let that wagging tail, incessant drooling, and pitiful stare convince you. Choose his food with care.

There are numerous products on the market aimed at each developmental stage from puppyhood to the senior years. Product claims include a shinier coat, whiter teeth, and even fresher breath for your dog. While the discerning dog food buyer can be confident that most products are healthy and nutritious, some are, indeed, more suited to specific needs, including age, health concerns, weight, and even taste preferences.

Don't be fooled by the price of dog food; more expensive isn't always better. Look for a stamp of approval by regulatory agencies dedicated to pet food and livestock feed, such as the National Research Council (NRC) or the Association of American Feed Control Officials (AAFCO).

The NRC requires that all dog foods contain certain essential elements. However, each dog food company usually has its own idea of how much of each element should be included, as well as the form the elements may take. For example, there's a choice between soy or animal protein. (See the box on page 133.)

Puppies' diets especially need to be monitored properly. Their largest growth spurt occurs from birth until the end of their first year. It's critical to monitor the nutrition they receive at this time. An overabundance of some of the nutrition staples, for example, may exacerbate any symptoms of hip dysplasia and other joint diseases in breeds that are susceptible to these conditions. Get your vet's recommendation on when to switch your pup's diet from milk or milk replacer to puppy food, which should be done before he leaves his mother, then from puppy food to adult dog food.

Health problems will affect how much, how often, and what type of food a dog or puppy should eat. Generally, dogs that are ill or stressed in any way need protein. Again, your vet is the best source of information here. She may even recommend food that is only available through her clinic. A reputable vet will not try to push these sometimes more expensive foods purely out of self-interest—choose your vet carefully *(page 158)*. If you find that your dog rejects this new, sometimes drier food, introduce it gradually. Mix a

Overleaf: Pekingese

bit of the old with the new, slowly increasing the amount of new food until the switch is complete.

It's best to find a few dog food brands or flavors that your dog likes and can digest, and rotate them over the long haul. Switching foods too often may wreak havoc on a dog's digestion, so when introducing the next brand in the meal rotation, do it gradually (as for vet-recommended food).

If Fido turns up his nose at dry or specialty diet food, try adding a bit of warm water—some brands even suggest this for a gravylike effect—or switch to canned. Canned food may sometimes have a strong odor to humans but most dogs love it. Refrigerate the leftovers promptly. Just microwave it on a plate for a few seconds before serving it from the fridge. If your dog is fussy, first check with your vet that he's not sick or allergic to the food. Otherwise, keep trying until you find a type that he likes.

Whatever you do, don't forget water. Your dog needs a steady supply of cold, clean water for cooling down and keeping internal bodily processes humming smoothly. Dogs who eat dry food will need plenty of water to add to the roughly 10 percent moisture found in dry food. Canned foods are already moist, so you won't need to be as vigilant in watching how much your dog drinks. Make sure you replenish the water supply as needed, and that you keep your pet's bowls clean to prevent residue buildup.

Some people let their dogs drink from the toilet, making sure they keep the bowl clean and free of toxic chemicals, such as those found in some toilet-bowl fresheners. Your vet may discourage this practice, considering it unhygienic. If you don't want Rover to slurp a tall cool one in the bathroom, keep the lid down, since many dogs find this lure just too strong to resist. (Also watch out for nearby hazards, including most bathroom cleaners.)

Finally, never feed puppies or grown dogs food such as onions or chocolate, which are toxic to canines, or milk, which often causes digestive upsets or diarrhea. And remember, no matter how hungry your dog looks, resist the urge to offer table scraps. You'll only encourage begging behavior.

Dogs are social animals, born to coexist with others, so it's natural that they will want

A chunky dog, as the one shown here, may be cute and lovable, but this isn't healthy. If it doesn't help when you change or restrict your dog's diet, increase his activity level. If he has accompanying symptoms such as dry skin, lethargy, and a dull coat, he may have a problem that only a veterinarian can diagnose and treat. On the other hand, a dog that can't keep his weight up may be nervous, underfed, or suffering from a disease such as cancer or diabetes or a condition such as intestinal worms. If you can see his ribs, your dog probably has to gain some weight. A vet can suggest a plan to help do just that.

These Jack Russell terrier pups are being fed a good diet to ensure proper growth and future health. Your vet can recommend an appropriate diet for your dog's breed, size, age, and weight. To ensure that your puppy won't become an adult dog that guards his food bowl aggressively, remove the bowl for short periods when still half full *(page 103)*.

Opposite: For older dogs and larger breeds such as this Great Dane, straining the neck for food can take away from the enjoyment of the meal, especially if they suffer from a cervical vertebral malformation or disc disease. In addition, bending for food may further complicate the problems of a dog with an esophageal disorder, since the normal contractions that bring food to the stomach may be affected. If your dog has to bend his neck too much to get at the food, place his dish on a chair or stool. Or, you can buy a stand made for this purpose, which is adjustable to any height and holds food and water bowls securely. If you're using a chair or stool, don't use a breakable dish. In his eating fervor, your dog may knock it over.

to eat with their human family. Try feeding your dog before you sit down to eat, or feed him in a separate room. You may still have to work at training your friend not to beg *(page 119)*.

If you have more than one dog, and if they're not territorial and respect one another's rank in the group, it's sometimes simpler to let them all eat from a communal bowl. Otherwise, give each dog his own bowl, well separated from those of the other dogs, to keep the peace *(page 122)*. You may even have to feed them on different schedules.

How much should you feed your dog? Store-bought food brands usually provide a guide based on weight and height, and sometimes age. However, this is often too much for the average dog. Discuss portions with your vet, with the goal of keeping the dog's weight consistent.

How often you feed your dog can vary, too. Two meals per day (one in the morning, one in the early evening) ensure that he's never too hungry, and keep his blood-sugar levels fairly constant. However, puppies and sick dogs will need to be fed more often, sometimes up to four times per day.

OVER- AND UNDEREATERS

Wild canids are constantly in search of food for survival. Although you will never let your dog go hungry, his instinct to find food remains strong. So while it may be a nuisance when your dog is constantly pawing through garbage, sniffing at the table, or trying to cadge a snack, keep in mind that he's only following his survival instincts, and work to gently correct this behavior *(page 119)*.

Some dogs are allowed to eat all day; that is, food is left in their bowls at all times. This constant availability of food can lead to an overweight dog. Treats and snacks add up in calories, too. To check your dog's body condition, do the "rib test." Run your hands on either side of his body along his rib cage. You should be able to feel the outline of his ribs. With an overweight dog, you might not be able to make them out at all. On the other hand, if the ribs are too prominent, the dog is underweight. In either case, visit the vet to rule out any health problems: dogs may gain or lose weight with illness. You may see other symptoms; for example, dogs suffering from kidney problems will also urinate and drink more, and may vomit and be depressed.

Your vet can recommend dietary modifications or special foods, and for an overweight dog, probably an exercise program as well. It's vital to get a chubby pup back to a healthy weight, since overweight dogs are at risk of diabetes, heart problems, and cancer, among other things. Keep track of all the extra bits of food given outside of mealtimes, and be more stingy in doling out treats, or ask your vet for ideas on healthier alternatives. Underweight dogs, too, are at a higher risk for all types of illness, due to their reduced ability to fight infection, decreased reserves of fat and energy, and poor healing ability. These dogs may need dietary supplements to bring them back into the pink of health.

ELEMENTAL, MY DEAR

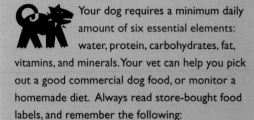

Your dog requires a minimum daily amount of six essential elements: water, protein, carbohydrates, fat, vitamins, and minerals. Your vet can help you pick out a good commercial dog food, or monitor a homemade diet. Always read store-bought food labels, and remember the following:

◆ Animal proteins are digested more easily than soy and other vegetable protein in general. You don't need to feed a dog as high a volume of food if it is easily digestible. The more digestible a food, the less stool will be produced. Keep in mind that a sick or stressed dog may need more protein.

◆ An unbalanced diet too rich in carbohydrates and/or fiber can cause constipation, bloating, and other digestive problems, as well as excessive elimination. Keep in mind that foods high in vegetable proteins are also high in carbohydrates.

◆ Fats keep skin and coats healthy and provide energy. Even an overweight dog needs a certain amount of fat in his diet.

◆ Rancidity can be a problem with food that has been sitting on the shelf for too long. Food treated with chemical preservatives such as BHA, BHT and ethoxyquin will last for up to eighteen months, whereas vitamin E and other natural preservatives will keep food nutritionally sound for six to eight months.

◆ A diet lacking in vitamins can lead to problems such as a weakened immune system, a greasy coat, bone disorders, thyroid problems, or behavioral changes, to name a few.

◆ Never give your dog mineral supplements unless prescribed by your veterinarian.

◆ Water keeps the bodily processes flowing. Make sure fresh, clean water is always available.

· PARLOR ·

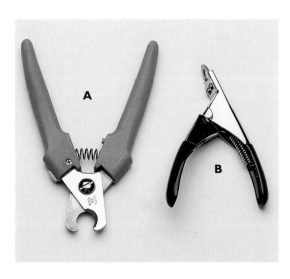

Clippers come in many sizes and shapes; the most important thing is to look for a tool that you can handle easily. Choose from a scissors-like clipper (A) or a guillotine-style clipper (B). Another alternative is an electric nail grinder, which is expensive but less likely to cut the sensitive vein (the quick) inside the nail.

Many—maybe even most—dogs love to roll around in the dirt or splash through puddles and mud, which means you'll likely have your work cut out for you when it comes to keeping your dog's coat clean and free of debris. But even for a dog who never runs into a burr or along a muddy trail, thorough and regular grooming is necessary for proper hygiene. As a by-product, you'll be able to keep track of any changes in your dog's skin, coat, teeth, ears, nails, and other assorted areas before problems have a chance to develop. Controlling fleas and other external parasites is another important part of your dog's coat care *(page 148)*. Rather than considering all this necessary maintenance as a chore, look at it as an excellent chance to bond with your best friend.

PAWDICURE

Unless your dog is gently introduced to the nail clippers, chances are he will fight you every time you try to trim his nails. Once he knows what to expect, you'll both be able to get through this necessary task unscathed. Let your dog get used to the nail tool's smell, look, and sound before you start.

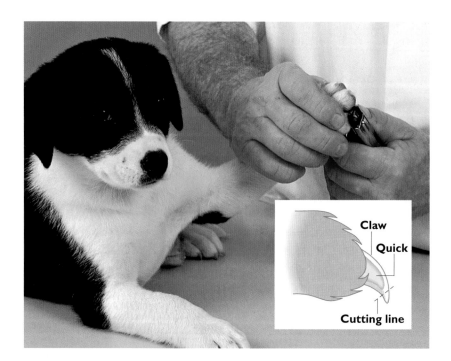

Claw
Quick
Cutting line

Trim a bit of nail at a time to avoid cutting into the quick *(inset)*. Watch your vet or groomer cut the nails the first time so you can see how it's done before attempting it yourself.

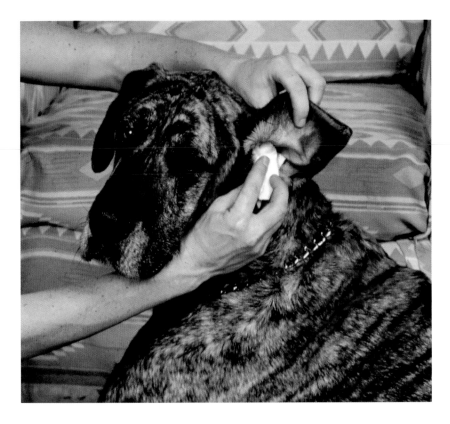

A gentle swabbing with a cotton ball dipped in mineral or baby oil, or a vet-approved ear-cleaning solution, keeps your dog's outer ear clean and dry. Dogs with floppy ears, such as this Great Dane, will need more frequent ear cleanings than their counterparts with upright ears because air doesn't circulate as freely; waxy and bacteria-laden debris tends to build up in the moist atmosphere under the flaps. Ear swabbing is doubly important if your floppy eared-dog loves to go in water.

Dirt easily gets trapped near and in the eyes of dogs with wrinkled faces or droopy lids, such as this bulldog, but any dog will need its eyes cleaned on occasion. Use a soft cloth dipped in warm water to gently clean the lids and around the eyes.

Because dogs have a sensitive vein, called a quick, that spans almost the entire length and width of each nail, clipping can be a tricky task. Consider yourself lucky if your dog has light-colored nails—the quick is in the pink area. Quicks in black nails are harder to find, so be extra careful to clip just a little at a time. (You can try holding a flashlight behind the paw to see the outline of the quick. In addition, a dark dot on the cut edge of the nail indicates the beginning of this nerve.) Have a clotting product on hand, available from your vet or a pet-supply store, in case you cut too deep. A styptic pencil, cornstarch, or soft bar soap will also stop the bleeding.

Choose a clipping tool that is sharp and easy to handle. When your dog is calm, hold one paw steady, speak in a soothing voice, and position the clippers over the top of the nail, just at the point where it starts to curl under. In one quick motion, squeeze the handle to clip the edge of the nail. Never clip too much at once, and take as much time as you need. For a hesitant dog, start with one nail, then slowly build up to four full paws over the course of a week or two. Don't forget to clip the dew claws, if your dog has them. You might at this point also trim the long tufts of hair growing between the toes, if your dog is a long-haired breed such as the keeshond. Finish off with a quick filing to remove any sharp or ragged edges.

Clip your dog's nails about once a month, depending on how quickly they grow. In the warmer months especially, his nails will be filed down slightly by

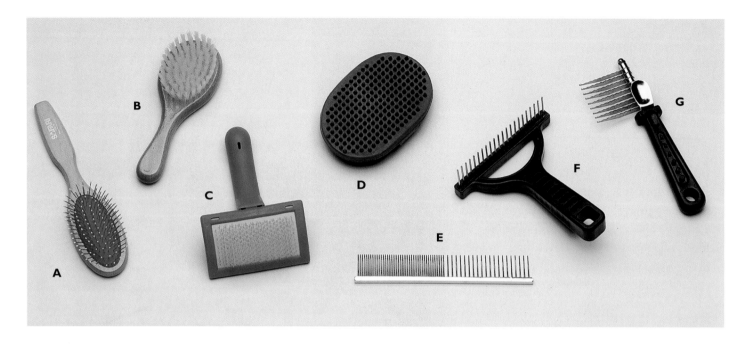

(A) The wire-pin brush is designed for breeds with medium to long or curly hair.

(B) Bristle brushes are used for any coat length. Long, widely spaced bristles are for long hair; short, closely spaced bristles are for short hair. Softer bristles are best for silky hair, and stiffer bristles for coarse hair.

(C) Remove mats and large amounts of dead hair with a slicker brush.

(D) Rubber curry brushes or mitts are ideal for brushing short, smooth coats.

(E) After brushing, use a fine-tooth comb for short or silky hair, or a wide-tooth comb for coarse hair.

(F) The rake is used to detangle and remove mats, as well as to pull off large amounts of hair during the shedding season.

(G) The mat splitter cuts out tangles.

friction from running on sidewalks and other rough surfaces, but this never replaces a good clipping. Neglected nails can break, crack, become ingrown, and affect your dog's mobility.

CLEANING EARS

Warm, moist, waxy and furry, the inside of a dog's ear is the perfect environment for all sorts of bacteria and infections. Keeping this area clean and dry is your best defense against unwanted settlements of mites, yeast, bacteria, and wax. An upright ear with little hair is easier to clean, and might even need less frequent care, because the circulating air will keep it naturally dry. Depending on the breed and the dog, you'll need to clean ears anywhere from daily to monthly. Each ear might also demand a different maintenance schedule. And for dogs with long ear hair growing out of the ear canal, such as poodles and terriers, carefully pull out as much of this fine hair as you can to make it easier to do the job properly. Place cornstarch on the hair to help you keep your grip on it, then clean the ear when you're done.

To clean the ear, use baby wipes, or dip a cotton ball or soft cloth in baby oil, mineral oil, or a cleaning solution recommended by your vet. Gently wipe the inside of the outer ear flap until there is no residue on the cloth (*page 135*). Never go in too deep, since this can compact the wax and damage the canal. Your vet can provide some commercial ear cleaner to squirt into your pooch's ear for canal cleaning. The dog will jump from this uncomfortable feeling, so hold him steady and massage the ear to help the cleaner flow into the ear canal. Clean out any dislodged debris with a cotton ball.

For a dog with a double coat, first brush against the direction of hair growth to make sure you're getting the wooly undercoat, which has a tendency to mat. Finish off by brushing the outercoat in the direction that the hair naturally falls.

If your dog's ear has a yeasty smell, is excessively waxy, or has a thick, dark-colored discharge, it may be infected. More clues include redness or swelling, and a dog that's always shaking his head or scratching his ear. A vet is the only person qualified to treat this, and no amount of cleaning on your part will do anything more than get rid of the continuing buildup.

EYES ON THE PRIZE

An inquisitive dog will poke his face into anything and everything, which can lead to injuries and foreign bodies lodging where they shouldn't. Keep a watch on Rover when you're playing with balls, frisbees, or other things that can hit him in the eyes. And do not let him stick his head out of the car window when you're driving. The risk of eye injury is too great.

You'll need to clean your dog's eyes of deposits of mucus, as well as accumulations of tears and other debris, using a soft cloth, cotton balls, or baby wipes *(page 135)*. Be especially vigilant with long-haired breeds such as the Shih Tzu and bichon frise, and those with droopy lids and excess skin around their eyes, such as the Chow Chow, boxer, and Pomeranian. Bacteria thrives in the dark, warm folds and soft furry hair, so take extra care when cleaning the eyes of these dogs. For dogs with long facial hair, clip the excess hair around the eyes with rounded-tip scissors. This keeps bacteria from forming around any buildup of tears and eye mucus.

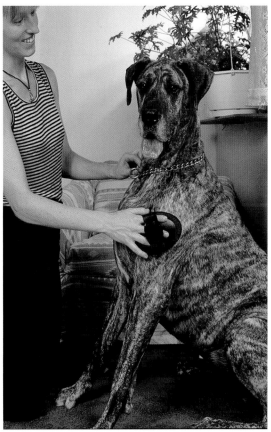

The short, smooth coat of breeds such as the Great Dane, Doberman, and boxer are easy to brush. A regular brushing with this rubber curry comb is about all this Dane needs to keep a healthy sheen. You can also perform a damp-cloth brushing, which picks up both excess hair and dirt.

A curly coat, such as that of this standard poodle, needs a lot of shaping and care. Dead hair can mat it down, so be sure to remove the old to make room for the new. A professional groomer is your best ally to help keep this type of coat clean, clipped, and healthy.

Keep a watch for signs of illness, including excess tears and mucus, swollen eyes, and incessant scratching. The culprit may be allergies, parasites, a cold, or eye trouble such as conjunctivitis, some of which can be caught from and passed on to your dog's canine playmates. If your pooch shows any of these signs, get to the vet quickly. And make sure your dog doesn't play with any other dog that displays these symptoms.

A SHINY COAT

Your dog may groom his own paws, legs, body, and even face, but he can never give himself the kind of care that you can with a thorough combing and brushing. There's no better way to keep your dog's skin and hair smooth and well-oiled, and his hair clean and tangle-free. A breeder, professional groomer, or vet can guide you in the choice of combs and brushes that best suit your dog's coat type *(page 136)*.

Be prepared to give some extra care to a dog with a double coat (both an outercoat made up of thick primary, or guard, hairs and an undercoat composed of soft secondary hairs), curly and thick hair, or long tresses. You'll have to work hardest in the spring and fall, usually high-shedding times.

In general, a quick daily brushing will clear loose hair and newly formed mats from the coat before there is a buildup, and a more intensive weekly grooming will keep your dog smooth—although the frequency and technique of grooming varies from breed to breed *(pages 180 to 203)*. The coat length and texture, and whether the dog has a single or double coat all affect grooming frequency. Even if two breeds have the same degree of shedding—high, for example—dogs with a longer coat, a double coat, or soft hair need to be brushed more often. This is because when shed, these types of coats tend to get tangled and matted more than would short hair, a single coat, or coarse fur.

To brush the whole body, section it off into smaller, more manageable parts. Start with the ears, then work to the head, neck, back, stomach, legs, hindquarters, and finish off with the tail. Be gentle with areas that have little or no fur; harsh brushing can be painful for your dog. If you need to work out mats or tangles, it often helps to have a second person hold your dog. Use a table or other raised area, always with a nonslip surface (one with a nonskid mat, for instance) at grooming time. Try not to use the floor, since it is too closely associated with playtime, which may mean a lively pooch who's too hard to handle.

Catch the shedding hair of very short, smooth coats, such as those on a Great Dane, Labrador retriever, boxer, or Rottweiler, by going with the grain with a comb, rubber grooming brush, or grooming mitt *(page 137, bottom)*. Or you can opt to use a damp facecloth, which will double as a surface-dirt cleaner. Longer-haired dogs with a single coat (an outercoat only) are brushed the same way, but with a slicker, wire-pin, or bristle brush. The narrowly spaced teeth of a flea comb will also work on finely textured short and medium-length coats.

For breeds with double coats, such as the collie, keeshond, Chow Chow, German shepherd, and Pomeranian, use a slicker or wire-pin brush. Make sure that the undercoat isn't overlooked, separating the coat to brush those hairs against the grain down to the skin *(page 137, top)*. Then brush the outercoat in the direction of hair growth.

Groom curly- and wiry-coated breeds, including poodles, schnauzers, and most soft terriers, with a wire-pin or slicker brush *(opposite)*. A special brush with rounded tips called a pin palm brush will smooth a terrier's wiry facial hair and legs. For breeds that don't shed, such as poodles, regular attention by a groomer will be required. Curly coats generally need a clipping about every six weeks. For a wiry coat, the groomer will use a painstaking technique called hand stripping to clear away all the dead hair that can mat this type of coat. This should be done every two to four months.

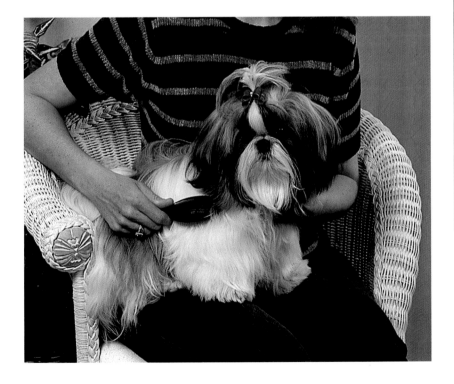

REMOVING MATS OR BURRS

 Knots, tangles, and burrs are par for the course during outdoor romps, especially those off the beaten path. Mats that pull on the skin and become even tighter when wet can form when a dog licks or scratches itself.

To remedy the problem, use a special tool, the mat splitter, to cut off matted and tangled clumps of fur and to remove burrs wedged into the undercoat. To avoid nicking the skin, only cut out a mat with scissors if you can slide a comb between the mat and the skin; then carefully cut over the comb. Otherwise, seek the help of a professional groomer.

A mat splitter, with its sharp-edged teeth, safely cuts out tangles and knots.

Shih Tzus, Afghan hounds, Yorkshire terriers, and other breeds with long coats require constant care to prevent tangles. For breeds with delicate coats, such as the Yorkie, go lightly with the brush to avoid damaging the hairs.

Dogs with long, silky coats, including the Afghan hound, Yorkshire terrier, and Maltese, need constant brushing to keep tangles from forming. Use a slicker or wire-pin brush to maintain the coat's smooth, beautiful appearance. Brushing gently, work from the skin outward against the grain, then finish off with strokes in the direction that the hair naturally falls (*page 139*). Because these coat types are long and sometimes hard to maintain, you might consider having your dog's coat professionally trimmed to a more manageable length. This will make it a lot easier to care for your pooch, although your dog may not look like the standard for his breed. You might find the extra maintenance (or professional groomer's fees) to be a worthwhile price to pay for a dog who conforms to the breed standard.

BATH TIME

Although many hunting dogs love to venture into the lake to retrieve a stick, or just to romp in the water, most dogs—like some small children—will act as if they are allergic to water when you try to give them a bath. Start your dog off early in the tub, train him to allow himself to be bathed, and reward him continuously while washing him, but remember that he, like most dogs, will probably be keeping an eye open for a way out.

Before you start the washing, brush your dog thoroughly and remove any mats. Use vet-prescribed ophthalmic mineral oil or a drop of eye ointment to protect his eyes, and cotton balls to gently plug his ears. Be prepared for the occasional wet shake-off during the bath, and dress accordingly. In warm

All you need for bathing your dog is a tub of warm soapy water, some dog shampoo, and the willingness to get wet. Work the shampoo into your dog's fur, making sure to massage it right to the skin. Always rinse him off completely afterward, comb him when he's still damp, then finish off with a good brushing when he's dry. A bath no more than once a month, or as needed, should keep Spot almost spotless.

weather, outside bathing will keep your bathroom floor from being soaked. Fill the tub or wash basin with warm water before you corral Spot. The water should reach to just past his hocks. (Or, for outside bathing, you can simply use a hose.) Using a sponge, lather him up using a dog shampoo which has been approved by your vet. If your dog has a specific problem, such as an extra-oily coat, you may need to use a medicated shampoo, available through your veterinarian. Work the shampoo into his fur from head to tail, paying special attention to known flea hangout spots such as the neck and in between the toes. Keep the warm, soapy water away from his mouth. If you know how, and if your vet has given you the go-ahead, you can empty the anal sacs at this point *(page 146)*. Rinse him off, then shampoo him again. Be sure to completely rinse his coat of any residue. Towel him dry.

Don't bathe him more than once every month or two, depending on the coat type. Bathing too often can lead to a dry, brittle coat and scaly, flaky skin. If your dog runs into a skunk, there are odor-removal products on the market, available at pet-supply stores, intended to get rid of the eye-tearingly strong odor, but you may still be left with the lingering scent for a few weeks.

PEARLY WHITES

Dogs aren't much different than humans when it comes to dental care. They need good tooth and gum cleaning, too, to ward off dental problems and gum disease. Plaque and tartar buildup and periodontal disease can even usher in heart, kidney, and other problems. Warning signs to report to your vet include red, bleeding, or receding gums, and persistent bad breath.

Ideally, dogs should have their teeth brushed daily, but at least twice a week may be more realistic. Brushing less often than every two to three days is ineffective, as this is how long it takes for plaque to harden. Make sure your pet is used to a dog toothbrush before you try to brush his teeth for the first time. If he seems nervous or uncomfortable with a toothbrush in his mouth, slowly desensitize him by rubbing your finger gently against his gums for a few minutes every day until he accepts this calmly. Praise him after each session. If he still rejects the brush, try doing the job with your fingers wrapped in gauze, or use a finger brush.

Never use your own toothpaste for your dog. He can't spit, and fluoride doesn't sit well in a dog's stomach. Use a special canine toothpaste supplied by your vet or available in pet-supply stores. If your dog absolutely won't let you use a toothbrush or your fingers, you may want to look into a tartar-combating oral spray. You might also want to ask your vet to prescribe a diet for your dog which helps to reduce tartar buildup. In any case, ensure that he gets a yearly dental exam at the vet. If your dog requires a dental cleaning, he'll be sedated, and each of his teeth will be scaled and polished. In addition, his gums will be thoroughly cleaned. As your dog ages, this yearly cleaning becomes extremely important in keeping him in the best of health.

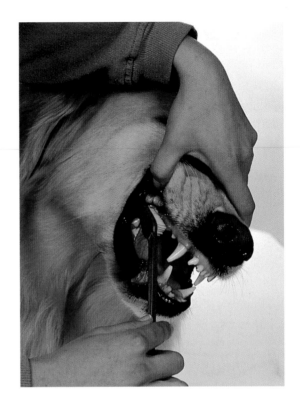

A great smile begins with healthy teeth and gums. Brush Rover's teeth using a gentle-bristled brush and toothpaste intended for dogs. You might have to use your finger wrapped in gauze or a small finger brush if he puts up too much of a fight. A brushing every few days should keep his mouth clear of tartar and bacteria buildup, although a daily brushing is even better.

COMMON HEALTH
·PROBLEMS·

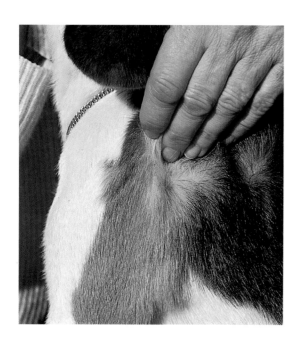

To monitor your dog's health, do a quick scan during your weekly grooming session. Check his whole body, from nose to tail, and look for signs of illness such as matted fur, swelling, or unusual discharge. Nose secretions should be clear, and his pulse—taken by pressing your fingers on the inside of his upper thigh—should fall within the range of 50 to 130 beats per second, depending on the breed (page 165). In addition, check for dehydration by twisting the skin of his shoulder. A healthy dog's skin will snap right back.

Just like a child, your dog will get into all sorts of trouble. Whereas a six-year-old may come home with a scraped knee or the occasional split lip, your dog's problems may be outdoor burrs, scratches from the tough cat in the neighborhood, and even the occasional bout of coughing or diarrhea. As you would with a child, pay close attention to your dog's health. Signs of trouble—such as changes in behavior and appearance—may be obvious, but they can also be very subtle.

Along with the medical diary you should consider keeping to show to your vet (page 158), being aware of your pooch's condition is your responsibility as dog caregiver. This means plenty of hands-on attention, as well as keeping your eyes and ears open. The sooner you report any changes in your dog's behavior or appearance to the vet, the sooner healing can begin. If not dealt with promptly, what could have been simply treated often blows up into a painful convalescence that is both time-consuming and costly.

Look for things such as odd moods and changes in eating and sleeping habits, activity level, and even posture. Note changes in your dog's bathroom schedule, as well as the amount and consistency of his waste. If his weight dips or soars from one month to the next, it may be time to give the vet a call.

It's also a good idea to have some knowledge of first aid (page 162). You can take a class or read up on it, but make sure that you feel confident about dealing with accidents and emergencies before you are put to the test—your best friend's life may depend on it.

BEHAVIORAL SIGNS

When your dog is really ill, you'll know it. He'll be lethargic, act depressed, and may not want to play or go for walks. However, slight problems may go unnoticed, since his canine mentality is such that he may act stoic and strong, hiding his affliction.

While some signs of illness are general and can be indicators of any number of dog diseases or acute problems (see chart opposite), other symptoms point to specific, common maladies. Mark everything you see and hear down in your dog's medical diary, and get to the vet as soon as you spot a problem. If you take your dog for a checkup every year and generally keep up with normal preventive care, you'll usually be able to nip any problems in the bud.

HEALTH PROBLEMS AT A GLANCE

Sign	Possible Causes or Conditions	What to Do
Abdominal pain or hardness	Blocked bladder, severe constipation, pregnancy, intestinal problems, pancreatitis, peritonitis	Consult vet immediately.
Changes in eating or drinking habits	Stress or a variety of disorders	Consult vet; if dog hasn't eaten for 24 hours, see vet as soon as possible.
Coughing	Allergies, upper and lower respiratory diseases, lung parasites, foreign bodies, heart disease, heartworm, abnormal windpipe	Consult vet within 24 hours, unless symptoms are severe or persistent. In this case call immediately.
Constipation	Diet, dehydration, intestinal blockage, eating bones or other hard substances	Add water or fiber (bran, pumpkin) to food, or add petroleum jelly to food with vet's okay. If condition persists, visit vet.
Dark residue or foul odor in ears, or dog shaking head or scratching ears	Ear mites, ear infection, inhalant and food allergies, foreign objects trapped in ear	Consult vet within 24 hours.
Diarrhea	Stress, change in diet, food allergy, intestinal infection or parasites, inflammatory bowel disease, parvovirus, coronavirus	If symptoms persist for more than a few days or are severe, accompanied by signs such as weakness, vomiting, or lethargy, consult vet at once.
Difficulty breathing; wheezing	Same as coughing, plus heart problems	Consult vet immediately.
Difficulty urinating; blood in urine	Urinary tract infection or irritation; stones, tumors	Consult vet immediately.
Excessive scratching or licking	Fleas. mites, skin disorders, allergies, wounds	Check for and eliminate fleas (page 148). Relieve itch with cold, wet towels. Otherwise, consult vet within 24 hours.
Excessive thirst and urination	Diabetes, kidney or hormone disorders, uterine infection, high blood calcium	Give plenty of water. Consult vet within 24 hours.
Foul breath	Dirty teeth, gum infection, abscesses, mouth tumors, foreign object between teeth	Brush dog's teeth (page 141), feed dry food. If symptoms are severe or persist, consult vet.
Inflamed eyes or eyelids	Eye infection, allergy, injury, glaucoma, corneal ulcers	Consult vet immediately.
Loss of appetite (anorexia)	Stress, gastrointestinal or other disorders	If dog hasn't eaten for 24 hours, see vet as soon as possible.
Loss of balance or coordination; weakness	Injuries, blood loss, brain or spinal trauma, poisoning, inner ear disease, tumors	Consult vet immediately.
Pale gums and mucous membranes	Anemia, heart disease, septic shock	Consult vet immediately.
Sneezing; runny nose or eyes	Cold or upper respiratory system infection, allergies, foreign object in nose, nasal mites or tumors	If problem persists for more than a few days or if dog stops eating, consult vet immediately.
Vomiting	Food allergies, intestinal problems, stress, or many other disorders	If symptoms persist for more than a day or are severe, accompanied by signs such as weakness, diarrhea, or lethargy, consult vet immediately.
Weight loss	May indicate many different disorders	Consult vet for diagnosis within 24 hours.

A change in your dog's temperature could be a sign of illness. To take this measurement, lubricate a rectal thermometer with petroleum jelly or mineral oil. With the dog lying on his side or standing, lift his tail and gently slide the thermometer into the anus—an inch or two for a small dog, up to half the thermometer's length for a larger dog. It will take about three minutes to get a proper reading. The normal range is 100 to 102.5°F (37.8 to 39.2°C), but these figures can vary slightly with recent physical activity, as well as excitement and stress levels. However, if your dog's temperature is above 104°F (40°C) or below 99.5°F (37.5°C), call your vet immediately. You could have an emergency on your hands.

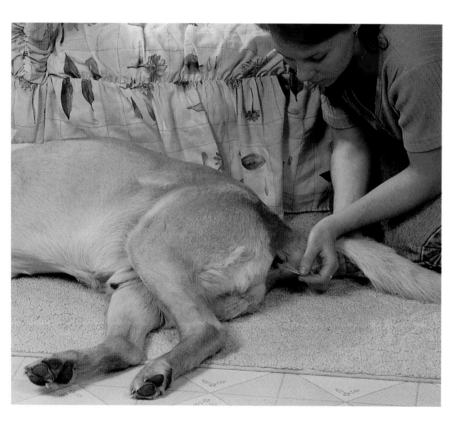

TIME TO CALL THE VET

There are many common canine ailments, some more likely to afflict certain breeds than others. Labrador retrievers, for example, are very susceptible to hip and elbow problems, skin allergies, and epilepsy, while Doberman pinschers are prone to hypothyroidism, blood clotting problems such as von Willebrand's disease, and a spine problem called Wobbler's syndrome. Although vaccinations will generally keep your dog safe from infectious diseases such as canine distemper, parvovirus, Lyme disease, and rabies, there are times when disease will get through your vigilant filtering and prevention system. You should always be on the lookout for signs of illness, no matter how many booster shots your dog gets.

Parvovirus damages your dog's intestinal lining, and is often fatal to young or unvaccinated dogs. Fever, weakness, a poor appetite, and depression, followed by vomiting and severe diarrhea are sure signs. If you see these symptoms, get the animal to the vet as soon as possible. Because "parvo" is picked up via the stool of an infected dog, keep your dog away from the feces of others. Dogs with coronavirus, a much less serious problem, will also show these symptoms, and might also have bloody stools.

The highly contagious and dangerous canine distemper virus shows up as a loss of appetite, nose and eye discharge, and such neurological problems as drooling, head shaking, and even seizures. In addition, look for hard skin patches on the feet or nose. Even if it's caught early, your dog may never recover, so make sure he is vaccinated.

Coughing and hacking can be signs of viral or bacterial infection, most notably parainfluenza or bordetella, respectively. Kennel cough, a broad term given to these and other infectious respiratory diseases, is usually passed between boarded dogs, or in any area where dogs come into close contact. General signs of listlessness and poor appetite followed by a recurring cough should send you and the pooch straight to the vet.

Leptospirosis, a bacterial disease that affects the liver and kidneys, can be picked up from contaminated urine. Look for signs of depression, weakness, abdominal pain, vomiting, diarrhea, increased thirst and urination, and mouth and tongue ulcers. This is highly contagious, and can be transmitted to humans, too, so rush to the vet if your dog shows these signs.

ZOONOSES

 Conscientiously looking after your dog's health is the way to keep him around for a long time. But did you know that by helping him, you're also helping yourself? There are more than 200 known diseases, called zoonoses, that animals can pass to humans—and household pets are known to carry thirty or so of them. Of the domestic animals, dogs carry the fewest.

Anyone can get fleas, salmonella, and tapeworm parasites from an infected dog. Rabies can also be transmitted, though this is unlikely in areas where rabies shots for dogs are mandatory.

Most zoonoses are rare, relatively harmless, and easily treatable and preventable, although the elderly, the very young and the ill are more prone to catching them. Cleaning your dog's food and water dishes with soapy water twice a week can keep the Citrobacter freundii bug—a close cousin of the stomach-upsetting E. coli—away. In addition, throw away old toys and bones which can trap the Bacillus bacteria. The following tips will help keep these sometimes nasty bugs at bay:

Avoid kissing strange dogs—or letting them lick you—on the mouth, or near the nose, eyes, or ears

Clean all scratches or bites, no matter how small, with soap and water, then hydrogen peroxide, and call your doctor. Red streaks radiating from the cut could be blood poisoning.

Earth and sand are a haven for common canine intestinal parasites such as hookworms, so protect yourself in areas where dogs may eliminate—keep your shoes on at beaches and wear gardening gloves.

Generally, keeping yourself, your dog, and sleeping, eating, living and waste-ridding spaces clean is the best defense against zoonoses.

One last word: You can pass the flu, Salmonella, fungal infections and other problems to your dog—second-hand smoke can be harmful to him, too.

If you can't get your dog to eat a pill encased in a piece of cheese, you'll have to drop it down his throat. Grasp his muzzle with one hand, and pull his jaw open with the other. Try to place the pill as far back into his throat as you can, then gently force his mouth closed. Massage his mouth, working down to the throat, until you see his neck move in the familiar swallowing motion.

Diseases from fleas, ticks, and other external parasites *(page 148)* can range from mild to deadly. Lyme disease, carried by deer ticks, is characterized by inflamed and sore joints, and a general lethargy. Heartworm, a potentially fatal parasitic infection, is spread by mosquito bites and shows itself by such signs as breathing trouble and general weakness. A daily or monthly pill given during the mosquito season—up to eight months per year—will protect your dog against the disease. Some common internal parasites, such as roundworms, hookworms, whipworms, tapeworms, giardia, and coccidiosis, can cause vomiting, anal itching, diarrhea, and dehydration. In addition, itching and swelling can be due to external parasites such as fleas, ticks, mites and lice.

Your dog's heart is susceptible to many illnesses and stresses. Heart muscle disorders can be a problem in large dogs such as the Irish wolfhound and Newfoundland, as well as in the Doberman pinscher and boxer. Problems with heart valves can affect older dogs and are most likely to strike toy, miniature, and small breeds—among them the Chihuahua, poodle, and Yorkshire terrier. Signs of heart trouble include tiring easily during play or exercise; coughing; increased heart and breathing rates; abdominal swelling; cold extremities; pale gums; depression; poor appetite or weight loss; and fainting. You might notice just a few of these symptoms, and even these may appear gradually. It's best to note any change in behavior in your dog's medical diary. If you suspect a heart problem, get to your vet or emergency clinic immediately.

A common metabolic disease to watch out for is diabetes mellitus, especially in old, overweight, and chronically ill dogs. Look for signs such as excessive thirst and appetite, increased urination, weight loss, lethargy and depression, and a dull coat. Diabetic dogs are prone to eye problems such as cataracts, characterized by graying lenses and ever-whitening pupils. Some may vomit, appear nauseous, and have sweet-smelling breath. If caught in time, diabetes can be treated with regular exercise, insulin injections, and a vet-approved diet.

Epilepsy is one of the most common neurological diseases in dogs. Look for symptoms such as seizures, strange behavior, and a lack of coordination. Any signs of problems with your dog's nervous system should prompt a visit to the vet. There, your dog will undergo a series of tests. If the problem is caught in time, he can usually live a long and happy life with the aid of anti-epileptic drugs.

Orthopedic problems such as hip and elbow dysplasia and arthritis can hit most breeds at any time, although larger breeds are particularly at risk. Improper posture and gait, limping, trouble standing up, and painful hips are the tell-tale signs. Keeping your dog at his healthy weight, providing a warm, dry living environment, and giving him vet-approved antiarthritic medications or prescription drugs will help manage the condition.

If your dog's backside is sore, if he drags it along the ground in an action called "scooting," or if there is an inflammation or swelling, the problem may be his anal sacs. These glands, just inside his anus, hold a liquid substance that is secreted onto his feces, perhaps as a scent marker. You may have to learn how to empty them if your dog has difficulty doing so. Some dogs do have trouble doing this, either because of an improper diet, obesity, inactivity, or because the sacs have become impacted because the fluid inside has thickened. The enlarged glands can be uncomfortable, and may occasionally become infected and abscessed. If you or a groomer use too much force when emptying them, the impacted glands can rupture. Or, they may not be emptied completely, and your dog's symptoms will continue. A qualified vet can check the glands, and teach you how to empty them properly.

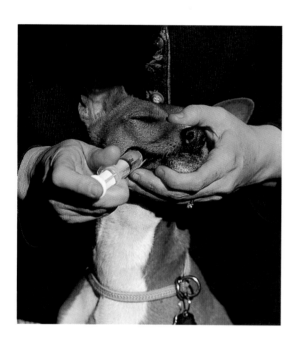

A few drops should do it, but not if you don't know how to give your dog his liquid medicine. Tilt his head slightly upward, then place a dropper or syringe filled with the correct amount of medicine behind the lip fold at the side of his mouth. Squeeze the liquid in a bit at a time to give your dog time to swallow.

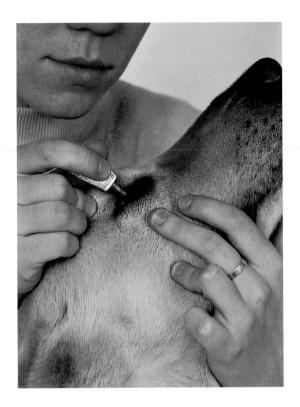

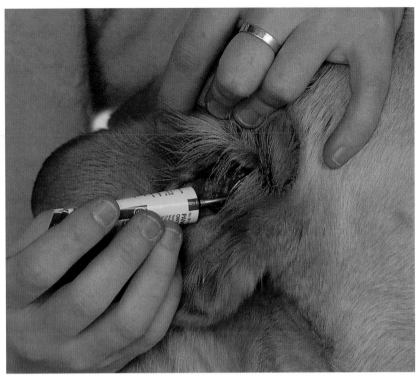

ON THE ROAD TO WELLVILLE

Most of the time, if a vaccination or pill did not prevent it, a simple schedule of vet-prescribed pills or drops will cure whatever ails your dog. (Always check with your vet before giving any over-the-counter medication. You'll avoid unforeseen reactions with your dog's diet or other medicine.) Your vet can show you how to give the pills or drops if you're not sure. Of course, getting your dog to agree to the treatment is another issue altogether. He might struggle or hide when you try to administer eye, ear, or nose drops, and he might resist pills by blocking access to his throat with his tongue. Many a tricked dog-lover has found a stray pill on the floor after he was sure Fido had taken his medicine.

To make things easier for both you and your dog, try to slip pills and capsules into food. If this doesn't work, you'll have to "pill" him *(page 145)*. For oral medicine *(opposite page)*, don't squirt in too many drops at a time or the dog might choke or have trouble swallowing. For ear and eye medication *(above)*, you may need a helper to keep your dog calm and still. Make sure the eyes and ears are clean and clear of debris and hair, and check for ear sores that would sting from an alcohol-containing solution.

Remember, even though you'll probably see a change in your dog's health and behavior after only a few days of treatment, don't ever stop the program midway. It will take the full prescribed time and medication amount to completely cure your dog of his illness and prevent relapses.

Above, left: When giving eye drops, use the buddy system if possible. Your friend can help keep your dog's raised head still while you gently roll back the upper eyelid and drop the medication in. Don't have any help? Kneel beside your sitting dog, then cup his chin and gently tilt his head so his nose points upwards. With one hand, pull his lower lid open. Use your other hand to pull back his top lid while you slowly squeeze out the medicine.

Above, right: Ears are often hit by bacteria and other invaders, so dog owners should get used to dropping medication inside. Sit your dog down, then kneel beside him as you pull up his ear flap. With the other hand, administer the medication, whether it's in a tube, syringe, or dropper, making sure it goes straight into the ear canal. Bring the ear flap back down, then rub it gently to work the medication inside. For erect ears, drop the medicine inside, gently fold the ear flap down to cover the ear, then rub the flap gently.

FIGHTING FLEAS &
· TICKS ·

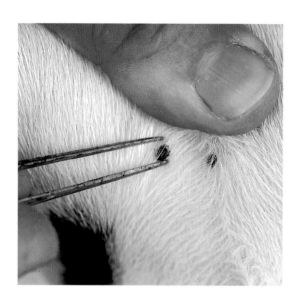

To remove a tick from your dog's skin, pull it straight out with tweezers, or put on rubber gloves and use your thumb and forefinger. Don't twist the tick. Its mouth, which is attached under the skin, might snap right off. If this happens, infection will follow unless you visit your vet to have the rest of the tick removed. Avoid touching any of the fluids from the wounded area or the tick itself, since it contains bacteria that can penetrate your skin. To prevent tick infestations, put a tick collar containing amitraz on your dog before he goes out, or apply anti-tick sprays or spot treatments.

After those much-loved outdoor frolics, dogs sometimes come in with more than muddy paws and burrs. Picking up external parasites is a risk whenever your dog ventures off the beaten path—or even into your backyard. Other animals can pass some of these pests on to your dog, too. And while they may seem like more of a nuisance than a health issue, some of these parasites can cause your dog serious harm.

External parasites are hard-coated insects and insectlike animals that live on the dog's skin, feeding off the blood, tissue, or skin itself. They include fleas, ticks, mites, lice, and flies. Fleas are wingless insects that leap onto your dog and suck his blood, sometimes transmitting other diseases such as typhus, or internal parasites such as tapeworms. An infestation can be fatal in some cases. The eggs laid on your dog's skin drop into his bedding, the carpet, floor cracks, and other places. The eggs develop into larvae, then pupae, maturing to adult fleas in a cycle that can take from two weeks to several months. The newly hatched fleas jump onto your dog, and the cycle begins again. Ticks are found in wooded and rural areas, attach themselves to your dog's skin, and cause swelling. They transmit illnesses such as Lyme disease as well as internal parasites. Mites live in the ear, under the outer layers of dead skin and on the hair follicles. If you see heavy black or dark red ear wax, itchy, dry patches, or hair loss, take your dog to the vet for treatment. Lice are not common in well-groomed and frequently-bathed dogs. Attached by the legs to their host, they survive on blood or skin scales, cause itching, and can carry other parasites, such as tapeworm larvae. Finally, insect bites can cause irritation, especially along the edges of erect ears, and some cause secondary problems; for instance, mosquitoes are responsible for deadly heartworm infestation (*page 145*).

STOP BUGGING ME

Prevention may be your best bet in the war against these pests. Keep your dog's fur clean, dry, and unmatted. At winter's end, look into flea- or tick-control products you can use before a problem arises. But whether you are warding off a potential infestation or dealing with an actual attack, ask your vet for the best product for your dog. It is important to use vet-approved products; in fact, some, such as the newer systemic and monthly topical flea-control treatments, are only available through vet clinics. Some flea-control products also kill ticks or lice.

FLEA CONTROL PRODUCTS

Some of the products available kill only adult fleas, while others are directed at only eggs and larvae. It's important to get at the larvae and eggs to prevent further infestations, but killing only these immature forms does nothing for the immediate problem. Remember that you have to get the bugs at all their stages—this may take several applications over the course of a month or two. Start with the least toxic formulas possible, especially with old, very young, or sick dogs. Never use two products in combination unless your vet gives the okay.

Product/Category	How/Where is it Applied?	Is it Toxic?	How Does it Work?
Flea comb	Groom dog, drop fleas into soapy water	No	Removes adult fleas and eggs
d-Limonene	Spray on dog, environment	No	Poisons adult fleas and larvae
Imidacloprid	Apply liquid to fur on dog's back	No	Poisons adult fleas
Lufenuron	Add liquid to food, or inject	No	Interferes with development of flea eggs and larvae
Insect growth regulators	Apply to dog, environment	Nontoxic to somewhat toxic	Interferes with development of flea eggs and larvae
Diatomaceous earth	Sprinkle in environment	Somewhat	Dehydrates adult fleas
Borax	Sprinkle in environment	Somewhat	Dehydrates adult fleas, eggs, and larvae
Fipronil	Apply liquid to fur on dog's back	Somewhat	Poisons adult fleas
Pyrethrins, permethrins	Apply to dog, environment	Yes	Poisons adult fleas and their larvae
Carbamates	Spray on dog, environment	Very	Poisons adult fleas and their larvae
Organophosphates	Spray on dog, environment	Very	Poisons adult fleas and their larvae

Application methods differ. Systemic products, often used as a preventive measure, enter your dog's bloodstream via a pill or an injection. Once the flea bites the dog and ingests blood, its eggs can't develop. However, if your dog already has fleas, you will need a secondary treatment for the live ones. There are topical treatments such as spot-ons, powders, rinses or dips, mousses, and sprays. Or, you can give your dog a bath using insecticidal shampoo, making sure to leave it on for the recommended time. Ask your vet to prescribe a combination treatment to keep your dog and home parasite-free: The various products act differently, either killing adults, and/or eggs and larvae. Keep in mind that if your area is prone to infestation, a product containing growth regulators that prevents hatching of eggs and larvae is a good idea. Aside from all these products, there are also flea combs, which remove some—but usually not all—fleas or lice; and flea collars, which are usually ineffective in the long run, and may cause dermatitis.

CLEANLINESS IS NEXT TO DOGLINESS

If you end up with an infestation, thorough cleaning helps. Wash bedding frequently (adding some flea shampoo), sprinkle desiccating products on rugs, and vacuum often. In addition, there are sprays for both indoors and the yard (especially shaded areas). You can also treat your house with a flea bomb or fogger, but this will require thoroughly cleaning any residue.

PREGNANCY &
· BIRTH ·

A whelping box like the one these Samoyeds are using will offer a soft, comfortable place for your dog to give birth; it will also serve as a safe location for the pups to spend their crucial first few weeks. Make sure the sides are low enough to allow the new mom to come and go as she pleases, but high enough to keep her pups from escaping. The shelf running around the inside of the box protects pups from accidentally being crushed against the side by their mother.

The sight of a litter of newborn puppies can make even the coldest heart melt. Bundled together, seeking warmth and food from their mother, and growing stronger every day, they are truly a marvel. Of course, if they were unexpected, their arrival may have a dark side. Because of the problem of dog overpopulation, millions of unwanted canines are euthanized in animal shelters in North America every year. To prevent unwanted pregnancies, have your dog sterilized *(page 160)*. If you don't, and you have a female dog, get ready for the joys and many responsibilities of helping your dog raise her young, from the first signs of pregnancy to finding good homes for the puppies (something to start thinking about as soon as you know she's expecting).

Beware of pseudocyesis, or false pregnancy. Although your dog may act and sometimes even look pregnant, she may not be. False pregnancy can occur at any time from six to ten weeks after estrus. Your dog's belly may look swollen, and her breasts may enlarge and secrete clear or brownish liquid. In addition to vomiting, diarrhea, cramps, and other physical changes, she may experience mood swings or show nesting and mothering behavior. If she is in extreme pain, your vet can prescribe medication. Spaying will ensure that a false pregnancy will not happen again.

A BUN IN THE OVEN?

If you catch your unspayed dog getting too friendly with the neighborhood stud, you probably have reason to suspect the impending pitter-patter of little paws. Take your dog to the vet as soon as you send the neighborhood lothario packing: A pregnancy can be terminated by a series of injections of female sex hormones starting within a day or two of the mating; your vet can inform you of the risks involved. Spaying is another option your vet might propose. If you prefer the wait-and-see method, the vet can detect a pregnancy by performing an ultrasound test as early as twenty-three days after conception, a simple blood test after about day twenty-seven, and, depending on your dog's body fat content and other conditions, a palpation exam after about day thirty. A radiograph after about day forty-five can confirm the pregnancy and provide a puppy count. Otherwise, unless you're aware of the changes your dog will undergo, you probably won't be able to detect signs of pregnancy yourself until she's at least five weeks along. With only about eight to nine weeks (fifty-seven to sixty-three days) to prepare for the newborns, there will be no time to spare.

Once a planned pregnancy has been confirmed, your vet will determine whether your dog is physically mature enough to sustain it. Usually any dog of two years of age or more is fine; if your dog is younger, you'll have to discuss your options with your vet. If your dog is on any medication—specifically, antibiotics and corticosteroids or other hormone-based products—your vet will likely stop prescribing it, since many can harm the unborn litter.

Physical changes you'll notice throughout your dog's pregnancy include a swollen abdomen and progressively larger and more pronounced nipples. Her body will be akin to a lightning rod in a hormonal storm, so expect to see mood and behavioral changes. In her last week, she'll begin to show signs of nesting behavior, typically by looking for a soft, warm, safe place to give birth. Have a whelping box ready *(right)* and make sure you show her where it is until she naturally goes to it.

Make sure your pregnant dog stays active during the gestation period: well-toned muscles will help her labor go more smoothly. If she needs to rest, she'll let you know. Likewise, maintain her grooming schedule so that she stays clean, calm, and happy. Take care to keep her warm after you bathe her, and try to not disturb her unborn brood, especially in the last few weeks of pregnancy. Be extremely gentle with her belly, applying no pressure there.

Keep her diet constant for the first three to six weeks of gestation, provided she's getting the usual high-protein kibble. Then gradually switch her to a vet-approved diet such as a naturally protein-heavy puppy food so that she has all the nutrients she needs to feed those growing prenatal pups. As the litter takes up more space in her body and crowds her stomach, she won't be able to eat very much at a time, so feed her smaller meals three or four times

WHELPING BOX SPECS

Your girl's delivery date is imminent, but are you prepared? A whelping box, where she can relax before and after the birth, may be all that you need. Build it, buy it, or use a cardboard box scaled to the soon-to-be-mother's size. Make sure there's plenty of room for your dog to comfortably stretch out with her litter of growing puppies. If you're using her traveling crate, just throw a blanket over the top so the new mom and her kids can have some privacy.

No matter what you use, sharp edges, toxic paints or substances, and sides so low that the inquisitive puppies can climb out are all no-nos. Place the box in a quiet area, line it with newspaper, and keep one of your dog's favorite blankets or toys inside to make her more comfortable.

To provide a warm spot for the puppies, set up a heat lamp over one corner of the box, adjusting it so the area underneath it is about 85°F (29.4°C). This way, you can place the newborns in the warmth, but the mother can move to a cooler area of the box if she desires.

Whether wild or domestic, newborn pups instinctively look for their mother's nipple. This gray wolf is nursing her cubs just outside the den. A whelping box takes the place of a den for domestic dogs, providing a sense of security during and after the pups' birth.

FOSTERING AN ORPHAN PUPPY OR LITTER

 The maternal instinct is one of the strongest in nature. Even so, a canine mother may abandon her litter as part of the natural selection process of weeding out sickly puppies, or due to a lack of bonding with her litter. In addition, a spoiled dog insecure about her rank in her human family may abandon her pups. In any of these cases, or if a mother dies during delivery, the pups will need someone to take over their care if they are to have a chance of survival.

Taking on the role of mother entails feeding them, washing them with a damp cloth, stimulating their digestive system by cleaning their belly and anus, and holding them often to socialize them. Keep them in an incubator warmed with a heating pad or bottle to between 85 and 90°F (29.4 and 32.2°C); slowly lower the temperature to 75°F (23.9°C) by the fourth week.

As long as you stay in close contact with your vet for advice, and the puppies receive all the inoculations and medical care they need, your orphaned litter should grow as strong as nature would have intended.

Hand-feeding orphan puppies every two to three hours will give them the start they deserve. Use a baby nursing bottle, syringe, or eye-dropper to feed them. Ask your vet to show you how it's done, and to help determine the amount and type of formula.

a day to ensure she's getting all the nutrition she needs. Nutritional supplements, unless vet-approved, are best avoided.

PUPPY TIME

When your dog starts to lie in her whelping box or to rearrange the newspaper and blankets almost as if she appears to be making sure everything is in order, the big event is only a week or two away. When there are only two or three days to go, she may eat less and discharge thick, clear mucus from her vagina. Within a day of the birth, your dog's temperature may drop below 99 degrees Fahrenheit (37.2°C). Take her temperature once or twice a day during the last week to catch this sometimes subtle sign. Be prepared for the impending births by having the following on hand: clean cloths and towels; a pair of scissors, and rubbing alcohol to disinfect them; dental floss or sturdy thread in case you have to tie off any umbilical cords; povidone-iodine; a lubricant such as petroleum jelly; and a baby's nose-suction bulb.

Your dog will go through three progressive stages during labor. During the first stage, the cervix dilates and the pups begin to move into position for delivery. This can last anywhere from six to twenty-four hours, during which time your dog will pant, shiver, and act restless. She may vomit, and her belly may begin to droop. She might be fearful and need your reassurance. During the second stage—the actual birthing process—she may lie on her side in her whelping box or other chosen area, or she may remain standing. She'll whine or groan as the contractions become more severe and frequent. Reassure her with kind words and a gentle hand, but make sure that only one or two people are with her during this time so she won't feel disturbed. It is a good idea to have a helper available, just in case something goes wrong. The birthing process will last anywhere from six to eight hours for a typical litter of four to six puppies, but a larger litter can take much longer.

Once you see the amniotic sac begin to emerge from her vagina, the second stage has begun, and birth is imminent. After her "water" breaks and a straw-colored liquid emerges, one pup should come out within minutes. The new mother will tear the protective amniotic wrapping and eat it. Then, while she licks her puppy clean and helps to stimulate his breathing and blood flow, she'll chew and eat the umbilical cord. The next pup will emerge in the same way, anywhere from fifteen minutes to two hours later. The third stage of labor is the afterbirth delivery. One placenta is expelled after the birth of each puppy, and the new mother will usually eat some or all of it. While your dog rests between pups, make sure that the newborns have access to the nutritive and antibody-containing colostrum that her nipples produce at this point. While she's giving birth, move the pups to a warm part of the whelping box, or place them in a box heated to 85 degrees Fahrenheit (29.4°C) with a hot water bottle. Hypothermia (low body temperature) or cold temperature shock is a leading cause of death in newborn puppies.

TIME TO STEP IN

Dogs have been giving birth, unassisted, for eons, and normally you should merely observe. However, sometimes having their human family members standing watch can be a matter of life or death for the mother, the pups, or both. If you notice that your dog is too busy with one pup to take care of the next one, you should assist her. If your dog shreds the umbilical cord too close to the puppy's navel, clamp or pinch the cord, tie it closed using dental floss or thread, then disinfect it with povidone-iodine. If the canine mother doesn't break the cord herself, tie it off about an inch from the pup's stomach, cut it, and disinfect the end. If a puppy isn't breathing or appears very weak, swaddle him in a towel and remove mucus from his nose and mouth with the suction bulb. Gently massage his chest, turning him over periodically to arouse him.

Some circumstances require an immediate call to the vet, if not an urgent visit. For example, if your dog goes into labor less than fifty-seven days into gestation, the pups may be too immature to survive. Another indication of possible trouble is an undelivered placenta or two, which can contribute to a serious postnatal infection for mom. More warning signs include a dog's straining to get her pups out; more than a two-hour delay between pups; dark green or bloody fluid passed before the first birth instead of afterward; more than thirty minutes between the breaking of the amniotic sac and delivery; and a puppy's head emerging during a contraction, then slipping back into the birth canal afterward.

WHAT'S NEXT?

Take the new family for a veterinary checkup as soon as you can. In some cases the vet may give the new mom an injection to clear the uterus and stimulate milk production. Stressed or overactive mothers may require extra food every day to keep up with the energy and nutritional demands of nursing; ask your vet about her needs. Also discuss the pups' inoculation schedule with the vet; normally they will need their first vaccinations, as well as another physical check-up, between six and eight weeks of age.

Weaning the young usually begins at four to five weeks of age; you can start giving them solid food at this point. By six weeks they should be mostly onto puppy food. At around eight-and-a-half weeks, the pups are ready to leave their mother. Unless you have a big house and yard plus plenty of time, you won't possibly be able to care for both your dog and all of her pups. Try to make sure the puppies will be going to permanent homes and not to someone taking a pup on a whim, or as a surprise gift, or just because their kids want it. Consider running through the guidelines for responsible pet ownership *(page 68)* with the person to get a feel for their commitment to a dog. And now may be the time to ask your vet how soon your dog can be spayed, so this litter will be her last.

This Jack Russell terrier pup is getting some all-important dog-human interaction. A newborn pup should only be handled minimally for the first two to three weeks, to avoid stress. However, after this period, gentle handling by humans is vital to a puppy's socialization process. When he is about three to four weeks old, the little fellow also needs to learn about new aspects in his environment. You can help him with this by moving the whelping box to another room. Then gradually put him on surfaces different than that in the box, and give him toys to discover; increase the time of exposure to these and other new things as he adapts to change.

CARING FOR
·AGING DOGS·

Since he's likely been a good friend, you won't begrudge the extra attention your older dog may need in his golden years. He may be thirstier, ask to go out more often, and be a little more moody, but these are all normal changes. For the most part, longstanding routines are maintained; the thirteen-year-old dachshund shown above still enjoys his walks. However, if you notice that your dog's behavior is becoming a bit erratic or his appearance is changing, don't just chalk it up to aging—consult your vet.

He doesn't jump as high, run as fast, or play for as long a time as he used to, but your four-footed pal still has the same twinkle in his eye that he always did. It's usually as he is easing into his twilight years—sometimes as much as half his life—that your dog will fit your lifestyle like a well-worn shoe. You will have long since settled into a comfortable schedule of walks, play, care, and general ease with one another. Truth be told, there is nothing more relaxing than sitting in a comfy chair on a sunny afternoon, book in hand and your favorite friend at your feet. Of course, these years also bring with them the biggest issues you'll have to face in regard to your dog: lifestyle changes to accommodate his slowing pace and medical problems he can live with—or can't.

THE GOLDEN YEARS

Even though your dog may be considered a canine senior citizen, you wouldn't necessarily know it to look at him. Depending on the breed, dogs slip into the golden years at anywhere from five to nine years of age. Smaller breeds, such as the West Highland white terrier, tend to live up to fourteen years or more, whereas some of the larger breeds, including the Great Dane and Irish wolfhound, often don't live beyond eight or nine years and begin to slow noticeably around the age of five. Generally, the smaller the breed, the longer its life span, with most of the in-between sizes enjoying a life span of between ten and twelve years. Senior citizen status is accorded to smaller breeds at about the age of nine, and to the larger breeds at about six years.

With aging comes a slowing metabolism, which often means fewer of those long wrestling and fetching sessions. This, coupled with your dog's tendency to store fat, may produce a pudgy pooch, so ask your vet to help you choose the right food. To keep your dog from further packing on the pounds, she may want to change his food to a higher-fiber, fat- and calorie-reduced "senior" formulation. High-protein foods may help your dog maintain his lean body mass. Kidney, liver and other problems that can accompany old age may require more specific diets. Your vet can also help you keep track of a weight change that may signal an illness.

Another way to keep your dog fit is to avoid letting his daily exercise slide, no matter how content he seems to be watching the world from the front window. Slow down your pace and shorten your walks, if need be, but don't forego activity altogether. To help him get his stiff, arthritic joints

Just because he can't get around as well as he used to doesn't mean that he has to stop doing the things he loves. With the help of a simple, textured wooden ramp or a store-bought model such as the one shown here, your dog can still climb aboard for those family rides you both love.

moving each morning, or to help ease the nagging pain of hip or elbow dysplasia, spend a few minutes gently massaging his joints. Even if you've never given a massage in your life, this gentle contact will be a boon to your dog's health and mood. If you're short on time, you might consider focusing on his ears and feet to give him a jump-start to a pain-free day: According to practitioners of dog acupuncture and massage, the ears and feet contain all the energy paths for the entire body (although such paths are scientifically unproven). As an added bonus, when you're massaging your dog, you'll be likely to notice any lumps, bumps, and skin and coat changes, all of which should be reported to your vet. Softer bedding and vet-approved vitamins might also soothe creaky joints.

A little compromise is to be expected. If you notice that your dog is having trouble hopping up to his favorite couch, either teach him to stay down, place a stool nearby to help him hoist himself up, or provide a soft pillow for him to lie on. Loading your older dog into the car can also become a problem. If he can't jump into the back of a high minivan, or even hop into the back seat of a car, use a strong plank of wood with a nonslip surface as a ramp to help him walk with dignity into his favorite cruising seat. Elevating his food dish to chest-height *(page 133)* is an especially good idea with an older dog, since bending only contributes to more pain and neck-strain problems. Do all that you can to ensure that his comfortable daily routine doesn't change too much. Dogs don't like to veer too far off their familiar course.

That distinguished gray beard, those white tufts between his toes, and his salt-and-pepper coat are other signs that your dog is getting along in years. However, don't let the gray fool you into thinking that he doesn't need as much grooming as he used to. Brush and clean him as always, using a more delicate touch if necessary. In addition, don't chalk up consistently bad breath to the normal aging woes. It may be a sign of illnesses such as liver disease, chronic indigestion, or stomach ulcers. Chronic halitosis can also be caused by periodontal disease, which can, itself, lead to other health problems, including heart, lung, and kidney disease. Keep up with your dog's dental and gum-care routine *(page 141)* and report consistent or recurring breath problems to your vet. As always, check your dog's ears, eyes, nose, coat, and full body, keeping alert for any changes that may signal illness.

Along with all the physical transformations he'll undergo, his sometimes strange behavior may make you wonder whether the wrong dog has followed you home. If he's lived to old age, he has earned the right to be ornery, sleepy, and a little moody. However, keep a watch on his behavior to rule out treatable and preventable diseases with behavioral symptoms, including diabetes—with its characteristic signs of increased thirst, urination, and weight change—and cognitive dysfunction (CD), the symptoms of which are forgetfulness, confusion, or sudden fearfulness, even with familiar people.

GERIATRIC PROBLEMS

You'll notice small, telltale signs of your dog's aging as time goes by. His eyesight and hearing may not be as sharp as when he was a rambunctious puppy. He'll tire more easily, be breathless on occasion, and may limp a bit in the morning. All this is natural, and neither you nor your vet will be able to turn back the clock. If the dog is experiencing pain, however, a vet can prescribe something to help. Happily, several age-related illnesses can be treated, if caught in time, so don't assume that every change in your dog's health is inevitable and natural. Many vets keep an eye open for the so-called "big five" age-onset illnesses: liver disease, kidney disease, diabetes, heart disease, and the ubiquitous cancer. A yearly series of "geriatric screening" tests—including liver, kidney, protein, and blood sugar checks—beginning when your dog is six years old, will help your vet to catch any problems before they become unmanageable. Blood tests can often catch diabetes and kidney and liver disease before there are even any symptoms. A stethoscope, electrocardiogram or X-rays can help your vet check for heart disease, and simply checking your dog for lumps is the standard first route to detecting cancer.

Although most types of heart disease can't be prevented, they can often be controlled with drugs, diet, nutritional supplements and exercise. Symptoms to watch for include excessive panting, coughing, and fainting. Diabetes can also be controlled with drugs, diet, and exercise. Symptoms of diabetes include diminished eyesight and a change in eye appearance and color.

Loss of appetite, depression, and increased volume and frequency of urination and voiding may be signs of kidney failure. Results from blood and urine tests, sometimes x-rays or ultrasound tests, and possibly a biopsy or exploratory surgery can help your vet determine the treatment plan.

If your dog is suffering from liver disease, the whites of his eyes may be yellow, his urine may be darker, and he'll be weak and lethargic. He may also eat less and drink more. To diagnose liver problems, your vet will need to do a complete medical work-up.

Biopsies from lumps and tumors will determine whether or not they are malignant. Some of these come simply as a result of age. If it's cancer, your vet will most likely remove the lump and/or local lymph nodes, or may start a course of radiation, chemotherapy, or hormone therapy. But you and your vet need to decide together whether the treatment and degree of improvement to your dog's quality of life is worth the stress he may undergo.

Arthritis and other joint inflammations may be a normal part of growing old, but the pain and soreness can be controlled with vet-prescribed medication. Don't take it upon yourself to medicate your dog. Aspirin and other drugs that can relieve these symptoms in humans can irritate a dog's stomach and lead to ulcers in the intestines or kidney disease.

SAYING GOODBYE

Nobody likes to think about the inevitable death of their trusted friend. If he's lucky, your dog will have lived a full, long life and will pass away quietly in his favorite area while he dreams. However, his human family might have to come to the painful decision of putting a loving dog to sleep to spare him needless suffering and loss of dignity. It all comes down to quality of life. If your dog is obviously not happy—and by this time, you'll be able to read his moods as well as he can read yours—and if his suffering cannot be helped by medical intervention, it's time for you to step in. This may the most selfless decision you can make on his behalf, owed to him for his many years of devotion and love.

Your vet will likely let you comfort your dog as she injects anesthetic into his vein. As you soothe and hold your friend, he will fall into a deep, painless sleep, then die peacefully.

Dealing with the loss of a dog is just as traumatic as dealing with the death of any loved one. It's important for children, especially, to have the chance to say goodbye when your dog is ill. Afterwards, you can have a memorial ceremony, scatter your dog's ashes in his favorite spot, or bury him in a pet cemetery or in your backyard, if zoning laws allow this. It may be a while before the family will be able to talk about your lost member, but in time it will be pleasant—and therapeutic—to share stories of how much your dog meant to you.

Although dogs generally slow down as they age, they might still occasionally play like the puppies they once were. This thirteen-year-old mixed-breed golden retriever still loves to fetch sticks during walks.

VISITING A
· VET ·

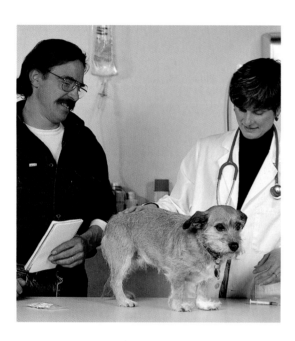

A qualified veterinarian will understand that your dog may have a few fears to overcome during his visit. Signs of a nervous pooch include flattened and pulled-back ears, tail tucked between the legs, and lowered head. Soothe your dog by talking calmly to him. If you have been keeping a medical journal, have it on hand when you visit so you can give the vet any necessary details about your dog.

Every person should have a yearly medical exam, and the same goes for your family pet. Vaccinations, blood tests, and physical checkups are important for good health throughout your dog's life. Daily maintenance measures are up to you, but only a qualified vet can tell if those under-the-weather blahs will require any treatment other than a hearty tummy rub.

PAYING A VISIT

Your dog's yearly vet checkup will include a complete physical exam, with a visual and tactile investigation of your dog's head, body, and tail, and all his assorted cavities. Because even the most cooperative dog might not readily go along with a regular tooth- and gum-brushing, a yearly cleaning by your vet may be in order. Like you, your dog can lose his teeth due to decay and neglect. It's a good idea to keep a dated medical diary not only of the procedures and vaccinations your dog receives at the vet, but also of notes on such things as your pet's elimination habits and any physical changes or unusual occurrences. Keep track of small shifts in his behavior, including urinary marking habits and mood swings, along with modifications in diet and other routines. Take the notebook when you visit the vet. These seemingly unrelated events might help explain results of your dog's medical tests. Also, if you're traveling, or if you need to change vets, it's good to have such a journal to provide a complete medical history.

Choose a vet who is calm, compassionate, and willing to explain all the procedures your dog undergoes. Don't be afraid to shop around to find a vet with whom both you and your dog feel comfortable *(page 88)*. For your convenience, choose a clinic with qualified staff and facilities to undertake surgery and perform procedures requiring anaesthesia, such as teeth cleaning. Because of the risk general anesthesia poses—especially for very old, very young, or very ill dogs—your vet will likely suggest a few tests, including a chest x-ray and a lab exam of blood and urine, before your dog is anesthetized. During the procedure he might need an IV drip; antibiotics may be necessary before and after. Ask your vet to outline the procedure for you.

Your vet may refer you to a specialist—veterinarians who have completed advanced studies in specialties such as surgery, internal medicine, and emergency and critical care—if your dog must undergo a procedure that requires more precise knowledge and experience.

You might want to look into medical insurance for your pet. Research the various insurance plans available, noting exactly what they cover, and weigh the cost benefit against that of using your credit card or having a special savings account for unforeseeable medical expenses that may come with caring for a dog.

A BOOSTER FOR BUSTER

Unless your dog lives in a secure room closed off to all people, animals, and even circulating air, he will no doubt come into contact with communicable diseases, bugs, and other assorted nasty things. Vaccinations and booster shots will help to keep him safe. Yearly vaccinations include those for the more common fatal diseases, including distemper. All-in-one shots will generally protect your pooch from distemper, hepatitis, parainfluenza, and parvovirus. Depending on local risks, protection from coronavirus, leptospirosis and Lyme disease may also be included. Also ask your vet about heartworm testing and prevention *(page 145)*. A rabies shot will be needed every few years. If Fido is boarded often, make sure he gets a preventative shot for kennel cough.

Some vets suggest testing your dog's immunity to distemper and hepatitis, among other diseases, and vaccinating only those at high risk for these illnesses. The cost of these tests can sometimes be higher than those of the shot, but there are some indications that too many vaccinations can over-stimulate your dog's immune system, which can sometimes lead to an increase in immune-related diseases.

Some vaccinations may have side effects, including nausea, vomiting, hives, and breathing trouble. Don't leave the vet's office right after your dog gets his shot, since most allergic reactions will occur within fifteen minutes. Keep a close watch on him for a few days after, and be careful when touching the tender areas where the shot was given. Inform the vet of any lumps, changes in skin color, or changes in behavior that persist for more than a few days.

ALTERNATIVE VET CARE

 More and more dog lovers are bringing in their pooches for alternative therapy, including homeopathy, acupuncture, massage, hydrotherapy, and even aromatherapy, when standard treatments don't help.

Stubborn chronic injuries and mysterious ailments are often tackled by alternative care. While controversial and unproven as long-term cures, these treatments are nevertheless carried off without drugs (and medications' potentially harmful side-effects), harsh treatments, or hospital stays. More and more vets now practice some form of alternative care, but if the practitioner you choose is not a qualified vet, make sure he is regulated, and either has adequate veterinary knowledge or consults with a vet.

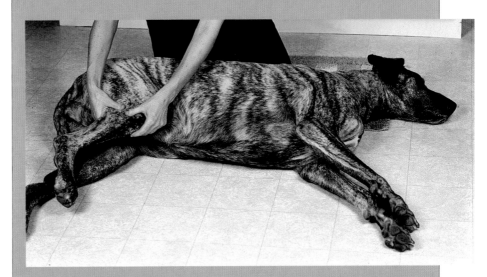

This Great Dane obviously finds massage soothing. Many people are turning to alternatives to veterinary care to ward off and treat illness in their dogs.

·STERILIZATION·

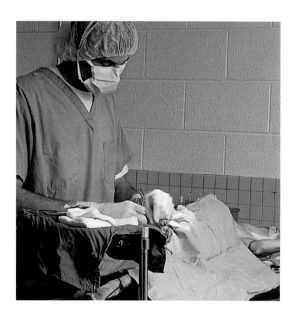

Help to control dog overpopulation with a simple sterilization procedure. Neutering, or castration (for males) or spaying (for females) is performed under general anesthesia. Your vet can inform you of any possible side effects of the anesthetic, as well as the facilities available in case of emergency. When your dog leaves the clinic, some post-operative medication may be prescribed. Follow your vet's advice regarding size and frequency of meals. Usually an absolute minimum of exercise is recommended for the first few days back home.

All puppies are adorable, but wagging tails and soulful eyes won't keep a surprise litter from ending up at the Humane Society. Each year, millions of unwanted dogs and puppies are euthanized. Unless you're a knowledgeable breeder continuing healthy, superior bloodlines, why bring more puppies into the world when there aren't enough permanent homes for them all? To avoid adding your dog's newborns to the long list of condemned animals, sterilize him or her as soon as you can. Not only will you decrease the dog's risk for certain diseases, including some types of cancer, and curb the sometimes uncontrollable behavior that a dog can exhibit, but having your dog "fixed" will benefit the entire dog population.

Other than the risks associated with any surgical procedure, there is no reason not to sterilize your pet. There's no truth to rumors of dogs becoming old, lazy, overweight, and apathetic after sterilization. This only occurs if you stop exercising and playing with him, or if you overfeed him.

Ovariohysterectomy, the most common spaying technique for female dogs, removes uterus and ovaries, as well as most of the cervix. A tubal ligation, not standard practice, ties the fallopian tubes so the ova can't enter the uterus, while another rare procedure, the hysterectomy, removes the uterus alone. Only the ovariohysterectomy decreases the incidence of mammary tumors and prevents the twice-yearly heat cycle. It should be performed anytime after a dog's fifth or six month, preferably before the first heat. Some vets will sterilize younger dogs. Discuss the pros and cons with your vet.

Neutering, or castrating, a male dog entails removing the testicles. Although this is encouraged before your dog is six months old, for population control, it prevents the development of his secondary sexual characteristics, including larger muscles. It also impedes development of his sexual responsiveness, a fact you'll appreciate if you've ever had to pry a dog off your leg. In general, the earlier your dog is neutered, the better, as he will be less likely to show hormonally-induced traits such as territoriality and overtly sexual and aggressive behavior. After puberty, castration can lessen these tendencies if they are already present. A vasectomy will not affect Rover's behavior; it stops sperm production without curbing the desire to breed.

Obviously, any diseases associated with the removed organs are avoided, and the risk for other, related ailments are significantly reduced. As a bonus, your sterilized dog's life also tends to be longer than that of non-fixed dogs.

THE PROCEDURE

While sterilizing your pet is a very safe, very common procedure, there are a few things you'll have to keep in mind.

Your vet will no doubt advise that you not feed your dog for at least eight hours before surgery. The dog will be given pre-anesthetic drugs to reduce anxiety, and to get his body ready for the general anesthetic. With an IV hooked up to his shaved forelimb or hindlimb, the sedative will allow the vet to insert a tube through your dog's mouth into his windpipe to administer gas anesthetic and oxygen.

For females, the ovaries, uterus, and most of the cervix are removed through an incision in the abdomen. For male castration, the vet makes an incision in or near the scrotum, and removes the testes.

Your dog will awaken soon after the stitches are tied. He might be in pain, so make sure your vet has prescribed painkillers. Never decide on your own to use human painkillers. They can be toxic.

Usually you can bring males home a few hours after the surgery, but females will often be kept overnight. Both will be back to normal in one or two days.

Monitor your dog's health, and watch for fever, swelling, vomiting, or discharge. Keep the affected area clean and dry. Usually you can begin feeding small meals and water a day after surgery.

Male
Kidney
Bladder
Testicles

Left: Neutering your male dog is a simple procedure. With common castration, the testes are removed through an incision in or near the scrotum. Give your dog an Elizabethan neck collar if he tries to remove the stitches himself. The vet will remove them about a week later.

Female
Kidney
Ovary
Uterus
Bladder

Right: An ovariohysterectomy removes the uterus, ovaries and most of the cervix of the female through an incision in the lower abdomen. A simple hysterectomy only removes the uterus, and won't prevent your dog from showing signs of heat. A tubal ligation keeps the ovum from entering your dog's uterus, but won't stop her from menstruating. Try to keep the dog calm and quiet for the first few days after surgery, until her incision has healed. Again, an Elizabethan collar may be useful if the dog is biting at her stitches before the vet is ready to remove them.

EMERGENCY
·CARE·

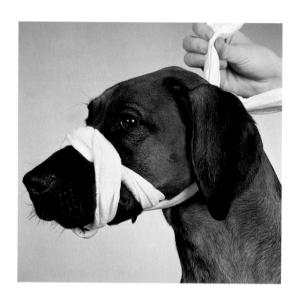

He's injured, scared, and snapping at you, but before you can approach him to help ease his pain, you'll have to get close enough. If you don't have a real muzzle on hand, use a handkerchief, a strip of cloth or gauze, a leash, or a length of heavy cord or rope to serve as a makeshift muzzle. Loop the material, approach him gently from behind, then slip it down as far as you can over your dog's nose. Draw it tightly closed, knot it under his chin, then tie the end pieces again at the back at his head. Don't worry about cutting off his breathing—his bony nose protects his nasal airway.

Dogs will be dogs, and with this special status comes a world of play, activity, and fun. Sometimes, however, sprains, breaks and even more serious accidents can befall a dog, or sudden serious health conditions can arise. Your responsibility as his human companion is to try to prevent problems while being prepared for every eventuality. Your immediate actions can mean the difference between life and death for your pet. First aid, as its name implies, is the initial treatment your dog will receive after the incident, to stabilize or comfort him. Usually your goal is to get him to the vet as quickly as possible, or at least to speak to a vet for advice. Post the numbers of your vet and the nearest twenty-four-hour emergency clinic in a handy spot, and also make sure you have the number of an animal poison control center. Keep a well-stocked first-aid kit *(page 205)* in an easily accessible place. Have an ice pack and extra bandages on hand, as well as larger items such as a board or blanket or towel that can double for a stretcher. Consider taking a course in canine first-aid and CPR to prepare yourself; CPR should only be administered if you know exactly what you're doing.

If you live in an area with frequent tornadoes or the potential for other natural disasters, it helps to have a well-thought-out plan of action ready, including a detailed evacuation plan and extra survival supplies *(page 205)*. Ask your vet or local humane society about the emergency services available in your area.

WHEN THE UNTHINKABLE HAPPENS
Having your wits about you and staying calm are the two most important things to keep in mind in an emergency—your dog will pick up on your mood. Take a few seconds to regain your composure, if necessary.

Whether he was hit by a car, attacked by another animal, shocked by a loose electrical cord, or even if he has ingested poisonous antifreeze solution which has spilled onto the ground, calm your dog as best as you can. If he looks like he might bite because of his fear or pain, muzzle him *(left)*, then restrain him.

As soon as possible, compare your dog's vital signs to their normal figures. His resting heart rate should be anywhere from eighty to 120 beats per minute *(page 165)*, and he should take from ten to thirty breaths per minute, depending on his size, age, breed, and physical condition. Normal rectal temperature should range from 100 to 102.5 degrees Fahrenheit (37.8 to 39.2°C).

EMERGENCY CARE AT A GLANCE

Problem	First Aid Required
Bleeding (cut, scratch, animal bite)	Apply pressure to wound until bleeding stops, then bandage *(page 165)*. If bleeding does not stop, apply tourniquet to a bleeding limb or tail *(page 164)* and get to vet immediately. If a foreign object is lodged in body, do not remove it; wrap a bandage around it and seek immediate vet care. If dog bitten by animal of unknown rabies status, seek emergency vet care.
Blood in urine/straining to urinate	Seek veterinary care immediately.
Burn, chemical	Flush with cold water and soothe with cold compresses. Seek veterinary care immediately.
Burn, thermal	Apply cold water or cold compress, then disinfectant. Seek immediate veterinary attention to check lungs for damage from smoke.
Choking	Remove obstruction, being careful of bites *(page 167)*. If not breathing, apply artificial respiration only if you know how and seek veterinary care immediately.
Convulsions	Move harmful objects away from dog and restrain him gently with towel. Record all details, including what dog may have consumed prior. If seizure is longer than five minutes or repeated, seek veterinary care immediately. Otherwise, call vet for advice.
Electrocution/electrical burn	Turn off power or remove from source of electricity without making direct contact—use broomstick. Seek emergency veterinary attention.
Fracture	Immobilize limb with splint in certain circumstances *(page 164)* then place dog on makeshift stretcher *(page 166)*. If bleeding, apply gentle pressure *(page 164)*. Seek immediate veterinary care.
Frostbite (pale, cool skin)	Slowly rewarm affected area with heat of your hand, by applying warm compresses, or by immersing in warm water (102° to 104°F, or 38.9° to 40°C). Seek emergency veterinary care if any pain, swelling, discharge, or discoloration or if skin does not return to normal after twenty minutes. Otherwise, get to vet within twenty-four hours.
Hypothermia (decreased alertness, weak pulse, shallow breathing)	Slowly rewarm by wrapping in warm blanket and applying towel-covered hot-water bottle filled with warm water. Call vet if dog does not return to normal when warm.
Insect bite/sting (may have large facial swellings, impaired breathing)	Pull out insect stinger, if any. Apply cold compresses to swelling to relieve itch and swelling. Seek vet care, especially with signs of allergic reaction such as difficulty breathing.
Poisoning (salivation, excessive vomiting, grogginess, unconsciousness, convulsions)	Call poison control center or vet, having product container on hand if possible. Induce vomiting only if instructed to, administering syrup of ipecac in dose recommended. Monitor for shock *(page 164)*; if convulsing, provide gentle restraint. Seek emergency veterinary attention, bringing product container or sample of toxin with you.
Shock (lethargy, rapid breathing, weak pulse, low body temperature)	Keep warm *(page 164)*; seek emergency veterinary attention.
Trauma, major (fall, car accident)	Monitor for shock *(page 164)*, keep warm, immobilize *(page 166)*, and stop bleeding *(page 165)*. Seek emergency veterinary attention.

If he's too agitated for this measurement, check his ears, nose, and extremities. If they're cooler or warmer than usual, his body temperature is likely off.

If you spot any cuts, scrapes, or other signs of injury, take him to the vet. If he needs to be carried, do so carefully; if he needs to be immobilized, get a helper and pick him up on a makeshift stretcher *(page 166)*.

TRAUMA

Until a vet determines that your dog will be all right, treat every wound and accident as if it's an emergency. Even a scratch can later become infected, and a puncture wound can become abscessed if not treated properly and promptly. Clean a minor wound with water or hydrogen peroxide (3 percent), then swab with povidone-iodine, chlorhexidene, or another antiseptic solution. Clean the surrounding skin and hair with soap and water. You can then apply an ointment such as neomycin. Avoid touching any odd-looking wound, since it could be the result of a snake or other animal bite, with venom or bacteria still present. Instead, wear gloves to clean it, then cover it with a bandage and get to your vet as soon as possible.

For a bleeding wound, apply firm pressure to the site of the injury using gauze, then bandage it *(opposite page)*. You'll see a slight swelling near the injury if the bandage is too tight. When the bleeding stops or slows, lightly cover the wound, then take your dog to the vet. If the bleeding is too severe to be stanched in this way and the injury is on a limb or the tail, you can tie a piece of thick string or gauze into a tourniquet at a point just above the wound, closer to the heart. Loosen the tourniquet for a few minutes every twenty minutes or so to let blood flow into the injured area. Get your dog to the vet immediately, and watch for shock *(box, left)*, which can result from too much blood loss.

Localized pain, limping, or swelling may be signs of bruises, muscle strains or sprains. Seek help if an apparently minor problem doesn't remedy itself in a few days. For a more serious problem, you might need to immobilize the limb or area with a splint *(below)*. Symptoms such as limb deformity or deviations from a joint or bone's normal position may indicate a fracture; take your dog to the vet immediately, especially if he has a compound fracture, where bone pokes through the skin.

TRANSPORTING AN INJURED DOG

Before you can get your dog to the vet, you'll have to decide whether to immobilize him, carry him, or just help him into the car. Of course this will be determined by the size and weight of your dog, and the extent of his injuries. Whether you splint a dog's broken bone depends on the fracture. With a mobile fracture, where the extremity is dangling below the break, create a temporary splint by wrapping the limb gently and thickly with padding, such as cotton rolls, or even a disposable diaper. Place a piece of heavy cardboard along the whole limb on the side of the injury, then secure it with an elastic bandage. This will keep the limb still, prevent nerve damage and ensure that there is no further interruption to the blood flow. Do not attempt to splint or immobilize the leg if part of it is dangling but you are unsure of the location of the break. If you can't splint a compound fracture, simply place a clean bandage over the exposed bone areas.

Try to gently ease your dog onto a makeshift stretcher, or hoist your dog by yourself (page 166). A small or medium-sized dog can be placed in a box for easier carrying. Partial paralysis of the limbs, accompanied by little or no pain, can indicate a spinal fracture, so don't lift your dog in this case. Immobilize him on a board to prevent any further injury to the spinal cord. Try to keep your dog still and calm as you lift him to place him in the car; do so as carefully as possible to avoid causing pain. Call ahead so the vet is prepared for you.

EYE INJURIES

A dog's eyes are very delicate, and untreated conditions, foreign objects, and seemingly mild problems can develop into rather severe trouble. Any change that is still present after twenty-four hours should be seen by the vet. Unexplained swelling or excessive tearing, especially if in only one eye, could be signs of infection or a foreign object in the eye. Inspect your dog's eye under a strong light, gently pulling the lids open. If the corneal surface isn't smooth or there's something protruding from behind the third eyelid (the nictitating membrane), gently rub a moistened cotton ball along the inner lid surfaces and under the third eyelid. If your dog doesn't easily comply, hold his head steady, then try to wash out the object with tepid water or eyewash. If you can't remove it, or if the eye is still red and irritated after a few hours, take your dog to the vet clinic.

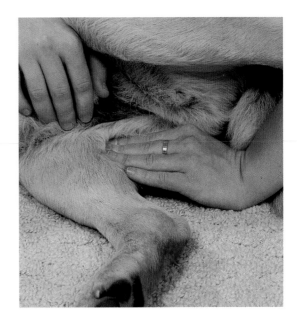

To determine your dog's pulse rate, place your fingers on the inside surface of his rear leg near the point where it meets his body. Count the beats for fifteen seconds, then multiply by four to get the pulse rate. It should range from eighty to 120 beats per minute, depending on the dog, breed and other factors. Or, for a quick check for a rapid or weak heartbeat, place your hand or fingertips against his chest, just behind the point of the elbow.

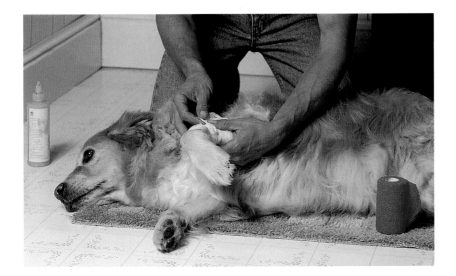

Once you've controlled the bleeding (page 164) clean the wounded area thoroughly, then cover it with gauze pads. For back, belly, and limb wounds, secure the dressing with cotton fabric placed on the wound, then wrap and tie it around his abdomen or limbs using a cotton sheet or strips, or stretch bandaging. Outer ear wounds are best left unbandaged, but those on the inside of floppy ear flaps should be covered. Fold ears on top of your dog's head to allow for good air circulation to the wound, then secure the ears in place with a bandage wrapped under his chin and tied on top of his head.

To transport a medium-sized dog without a blanket, stretcher or box, approach him from the side and cradle him with one arm underneath his rump and your hand supporting his chest. Reach your other arm around to support his head, with that hand holding the top of his front leg. Hoist him up and carry him close to your body. For a very small dog, reach around and place one hand just behind the point where his front legs attach to his body. Hold his lower quarters steady with this arm. Support his chest with your other hand. For a very large dog, you'll have to lift him by placing both hands underneath his belly, close to the point where his front and back legs meet the body.

Above, right: The best way to immobilize an injured dog to transport him to the vet is by creating a makeshift stretcher using a blanket, sheet, towel, or a flat, solid board. Gently ease your dog onto it, but don't force him to lie in any position, and don't tie or hold down a struggling dog. Small or medium-sized dogs can be carried in a box.

DIFFICULTY BREATHING

If your dog is having trouble breathing, there could be something lodged in his mouth or throat. Or, he could be having heart trouble, a reaction to an insect bite, or may even be in shock. To narrow in on the cause, do a quick check of his vital signs.

Anything that your dog can get his paws on will probably end up in his mouth and may get stuck in his throat. If he is calm enough, remove the obstruction *(opposite page)*. Alternatively, a canine-modified Heimlich maneuver might rid his airways of the blockage. To perform it, stand behind him and wrap your arms around his abdomen, just below his ribs. Apply a few quick compressions, or thump both sides of his chest simultaneously with cupped hands. Be careful about how much force you use; especially with a smaller dog, you might break his ribs.

BURNS

Burns, either from a chemical or thermal source, may not be visible under thick fur. Indeed, there may be no signs of skin damage, such as oozing pus or blisters, for up to three days. If you do suspect a burn or see symptoms, flush chemical burns with cold water; for a thermal burn, apply cold water or a cold compress, then disinfectant. Get your dog to the vet quickly.

THE GREAT OUTDOORS

When the weather is very hot or very cold, your dog's system may not be able to adjust. Heatstroke, *(box, opposite)* frostbite, or hypothermia may result. Decreased activity, shallow breathing, and pale, cool skin are signs of hypothermia. In such a case, the body temperature drops drastically, and your dog can't muster enough energy to even shiver in order to heat up. Warm him with towels or blankets, or use a hair dryer on its lowest setting. If he doesn't return to normal when his body temperature does, or if he seems confused, call your vet.

Paw pads, the tail tip and ears are particularly susceptible to frostbite and will turn white as a result. Bring your dog indoors or shelter him from the wind. Use your hands to rewarm the affected parts, and keep him warm with your body. If he is still listless and cold, slowly apply warm compresses. Don't use hot water or hot compresses, since reheating your dog too quickly can cause his blood vessels to dilate too rapidly, which can result in a dangerous drop in blood pressure. Always call your vet for advice.

SEIZURES

The symptoms of a seizure include uncontrollable and uncharacteristic shaking, loss of consciousness, and uncontrollable voiding. Don't try to restrain your dog. Gently wrap him in a blanket, protect his head, and wait for the seizure to pass. Make a note of anything he has recently consumed, his behavior before and after the seizure, and the length and intensity of the episode. If the seizure lasts more than a few minutes, or if there are repeated episodes, take your dog to the vet as soon as possible; otherwise, call your vet to describe the seizure and get advice on the course of treatment.

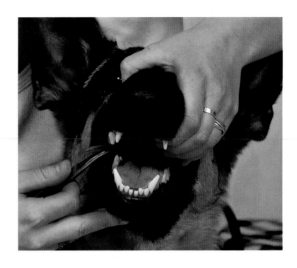

To remove an obstruction from a conscious dog's throat, you may have to use a pair of tweezers or needlenose pliers to reach the object. If possible, have a helper hold him steady while you hold his mouth open to do this. Remove the object carefully and report any swelling, hoarseness, or other problems to your vet. If your dog is unconscious, shift his head so his neck is extended, then open his mouth and pull his tongue out past his teeth. Sweep your fingers around his mouth and throat to search for the foreign body.

HOT DOGS

Have you ever left your dog in the car, with the windows rolled down slightly, to run into a store for "just a minute?" With the summer sun beating down on your car's roof, it only takes a few minutes for the temperature inside to rise to 120°F (49°C) or more—a climate that your dog can't tolerate. Since his only way of cooling down is by panting and allowing the moisture from the surface of his nose and paw pads to evaporate, he will likely overheat in no time at all. Signs of heatstroke include an increased pulse rate, heavy panting, anxiety, and confusion. If left untreated, he can lose consciousness. Cool him down with cold, water-soaked towels, and call your vet for guidance.

Don't ever leave your dog in a hot spot, enclosed area, or place where there's no air circulation. On hot summer days, make sure that he has access to plenty of shade and water, or an air-conditioned room.

DOG
BREEDS

· · ·

**"Money will buy you a pretty good dog,
but it won't buy the wag of his tail."**

HENRY WHEELER SHAW

PUREBRED
·BASICS·

The Indian Plains wolf, an extinct small Asian subspecies of the gray wolf, is thought to be the ancestor of a great number of domestic dog breeds, including many of the so-called primitive dogs, such as the basenji, the Canaan dog, and the dingo. Between ten and fifteen thousand years ago, this wolf's early domestic relatives accompanied migrating humans out of southwestern Asia and into the Middle East and Africa. Some are even thought to have headed east over land bridges, crossing what is now the Bering Strait into North America about eight thousand years ago. Genetic testing is currently under way to see if the resemblance between the primitive Carolina dog of the U.S. and Australia's dingo is more than skin deep.

Other dogs took a different path. Descendants of the larger European wolf migrated north, eventually spreading out across the Arctic tundra. Later, some turned south again into Russia, Alaska, Canada, and the Scandinavian countries, providing the root stock for Nordic spitz dogs such as the Siberian husky, the Alaskan Malamute, and the lesser known Greenland dog and various laika breeds.

EARLY BREEDS

The first breeders deliberately selected attributes for their domestic dogs. Now settled and likely beginning to farm, humans turned the first dogs, still physically similar to wolves, into guard dogs, hunting dogs, and herders. This changed the dog's shape, and the early breeders of the Far East, Middle East, and Asia soon had more to work with in terms of physical characteristics. Apart from size and shape, humans began to select a dog's coat type, color, strength, and speed. Before long, a variety of distinct types emerged.

By at least five thousand years ago, greyhound-type dogs began appearing in the Middle East. Skeletal remains and art decorating tombs of the pharaohs suggest that ancient Egyptians bred a hunting dog similar to today's pharaoh hound, sometimes called the oldest domestic dog breed. Before long, migration and trade placed dogs just about everywhere. More than two thousand years ago, Phoenician and Carthaginian traders are believed to have introduced some of the Egyptian hounds and mastiff-type dogs of Tibetan origin to many Mediterranean countries, and perhaps as far north as England.

The Romans and Assyrians were particularly fond of the early mastiffs, which they bred and employed not only as cattle dogs, but as dogs of war.

Primitive dogs such as the dingo are thought to be related to the Indian Plains wolf, a small, long-extinct subspecies of the gray wolf. Sometimes called feral or pariah dogs (Tamil for "outcast"), the dingo and others akin to it are simply domestic or partially domestic dogs that have reverted to a wild lifestyle. Having arrived in Australia with the early aboriginal settlers more than four thousand years ago, the dingo is now only very rarely kept as a pet.

Opposite: The Australian shepherd is a livestock dog that is not from "down under," more likely having originated in the Basque region of the Pyrenees Mountains between Spain and France. Its modern development occurred in California in the 1940s and 1950s. In 1991, the Australian shepherd was accorded official breed status by the American Kennel Club.

Overleaf: Airedale terriers

Although it is commonly cited as originating from Great Britain, the mastiff's origins are a bit of a mystery. It was most likely developed in southwestern Asia or Tibet by selectively breeding the largest available primitive dogs.

Fiercely aggressive and strong, they accompanied Roman armies onto the battlefield and, by extension, throughout Europe as well. Among others, the bulldog, the Great Dane, the boxer, and the St. Bernard are thought to have descended from these not-so-gentle giants.

MAKING FRIENDS

The skeletal remains of small dogs and the depiction of them in artwork suggest that apart from their interest in powerful fighting and guard dogs, the Romans were among the first to develop dogs for companionship. Small white dogs resembling today's Maltese can be found in Roman-era paintings. Other evidence from about the same time suggests Oriental origins for some of today's toy or companion breeds. The Pekingese, favored by Oriental royalty, is believed to be well over two thousand years old. The dogs were so small that they were carried inside the sleeves of the gowns of Chinese empresses.

Royalty and aristocrats continued to favor the companion breeds for many centuries. During the Middle Ages, the tiny Japanese chin was a favorite of the Japanese court. By the seventeenth century, breeds such as the King Charles spaniel and ancestors of the bichon frise, lowchen, and Maltese were living the royal high-life in England, France, Germany, and Spain.

Ever fond of hunting for sport, the ruling classes monopolized the hunting hounds. The peasantry was strictly prohibited from owning these dogs. The most formidable of the hunting hounds was the progenitor of today's Irish wolfhound. Literally intended to track, catch, and kill wolves, the ancestral breed and many others became extinct with the invention of firearms and the eventual disappearance of the lupine quarry. Around this same time, more specialized hunting dogs began appearing, the ancestors of today's pointers, spaniels, setters, and retrievers.

GETTING ORGANIZED

By the seventeenth century, hundreds of distinct dog breeds existed around the world. Selective breeding and dog fancy in general had become popular, if rather unorganized, pursuits. In 1735 Carolus Linnaeus, the father of lineage, listed only thirty-five different dog breeds in his *System of Nature*. It wasn't until the first dog shows and the rise of kennel clubs in the mid-1800s that more formal, or systematic, selective breeding began. With the seeds of breeding methodology and regulation finally (albeit loosely) planted, the development and showing of dogs with a pure and well-documented lineage expanded tremendously.

In the 1850s and 1860s, the first dog shows centered around sporting dogs, most notably hunting hounds such as the English foxhound, favored by fox-hunting aristocrats. In 1873 Britain's Kennel Club was formed,

Like the Pekingese, the Japanese chin, and several other small toy or companion dogs, the Shih Tzu can claim Asian royal origins. They may have first arrived in China from the Middle East or southwest Asia as far back as the first century. Later bred within the walls of the Imperial Palace of Peking, most of the Shih Tzu's development occurred during the Ming dynasty (1368-1644).

HYBRID DOGS

Because of their similar genetic makeup, domestic dogs can mate and produce hybrid offspring with some of their wild relations. Hybrid dogs are usually the result of deliberate pairings. Depending on your point of view, coyote or wolf genes are used to "reinvigorate" *canis lupus familiaris*, or domestic dog genes are introduced to produce a "softer" wild canid. Although their appearance can be breathtaking, it is both unfair and irresponsible to produce and confine hybrid dogs. While some may display the behavioral traits of their domestic relatives, many retain strong natural drives and instincts. Often, these animals are restless, constantly pacing the boundaries of their territories. More than half a million wolfdogs, or hybrid wolves, are said to exist in the U.S. alone. And whether they display the wolf's naturally timidity or not, they can be dangerous to humans. Keeping them as pets is strongly discouraged, and several U.S. states have banned them outright. Those who enjoy the look of wild canids should consider one of the Nordic breeds that closely resemble their wild cousins. Wolf lovers can choose a Siberian husky, Alaskan Malamute, Samoyed or keeshond. Fox fanciers may appreciate the Finnish spitz or the Japanese shiba inu.

Hybrid dogs also occur naturally, most often as coydogs like the Siberian husky-coyote male on the left. The more predictable nature of the purebred Siberian husky female on the right makes her the better pet.

COAT OF MANY COLORS

Selective breeding exploited what occurred in nature, creating dozens of coat colors and patterns from the wolf's original color range. Despite its name, the gray wolf injected a variety of colors into the canine gene pool. Apart from the wolf's gray, black, and even light-brown coats, its various camouflaged subspecies, such as the all-white arctic wolf and sandy-colored Indian Plains and red wolves, helped formulate a dog color range that now extends from pure white through brown, red, gray, and black, all in a variety of shades. Some shades have unusual names such as wheaten (pale yellow) or liver (bronze, cinnamon, or reddish brown). Lemon refers to the orange-brown patches commonly coupled with white in bicolor hounds, while blue is simply a combination of black and white hairs. These same colors also group together to create a number of different patterns by combining two or three colors or multi-colored individual hairs. For example, black and tan is a self-explanatory bicolor. Add white and it becomes a tricolor, also seen in many hounds. Harlequin means black or blue patches on white. Roan signifies a fine mixture of white hairs within another color. The brindle pattern creates tigerlike stripes by layering dark hairs over lighter ones.

This cocker spaniel sports a roan coat.

registering thousands of pedigreed dogs under forty recognized breeds. Similar institutions were soon founded in the rest of Europe and eventually in the United States, with the birth of the famous American Kennel Club (AKC) in 1883.

STANDARD PRACTICE

The various kennel clubs differed in many ways, but all insisted on one thing: dogs, armed with their pedigree paperwork, needed to be registered in order to compete in kennel club shows. For each recognized breed, the clubs developed a "standard," a precise, written ideal dictating both physical and behavioral archetypes, from head and body shape to gait and temperament. This greatly influenced the development of all breeds. Before there were kennel clubs and standards, appearance could—and would—vary wildly within a breed. But by the end of the nineteenth century, appearance conformity had surpassed purpose as the raison d'être of dog breeders. Soon dogs within each of the individual breeds began to look a lot alike.

Conformity certainly refined the breeds, but the inbreeding that led to consistent repetition sometimes brought forth a variety of hereditary health problems, such as hip dysplasia *(pages 144 to 146)*. Certain breeds are known to have other joint problems, while Dalmatians sometimes suffer from hereditary deafness. Other health problems have arisen from breeding to meet the standards themselves. Bulldogs, bred to have larger and larger heads, can suffer from a variety of respiratory problems caused by the flattening of their natural face shape. Many have had to be delivered by cesarean section as their large heads would not pass naturally through the birth canal.

CONTROVERSIAL PROCEDURES

Steeped in history and tradition, kennel clubs tend to be extremely conservative institutions. Wary of change, a few have balked at altering or eliminating the longstanding standards that call for some controversial surgical procedures—namely the docking of tails and cropping of ears—to be performed on certain breeds. Long ago, when many of the breeds in question still hunted, worked, or even fought for a living, their owners had their tails and ears cut short to prevent injury. Terriers purportedly even had their tails docked so that the stumps could serve as handles by which to pull the dogs from dens and burrows. Today, with a great many dogs retired into a life of companionship, there is little or no practical reason for such procedures. Tail and ear injuries are now extremely rare. In Australia and most of Europe, cropping and docking have been banned, and some organizations, including the American Veterinary Medical Association, are lobbying for similar legislation in North America, claiming that the procedures are cruel, unnecessary, and wholly cosmetic. Ear cropping can be particularly painful

MIXING IT UP

The world's most popular breed of dog is no breed at all. Mixed breeds, random-breds, mongrels, mutts, or curs, call them what you will, they make up the majority of the worldwide dog population. Rare is the country where dogs of mixed and usually unknown heritage do not outnumber their blue-blood, purebred relations. In true mixed breeds, the dog's ancestry is next to impossible to predict, although many people can't help but try to guess. That's part of the fun.

Because they're all related, all of the four hundred or so breeds are capable of interbreeding. The millions of mixed-breed dogs around the world are a testament to that. Unfortunately, these are the dogs you're likely to find at the local shelter or dog pound, often the result of accidental breeding between two unsterilized dogs.

Often used interchangeably, the terms "mixed breed" and "crossbreed" have slightly different meanings. Unlike mixes, crossbreeds have clear roots—often evident by looking back just one generation. Sometimes produced randomly, but most often planned by breeders, crossbreeds result from the mating or crossing of two dogs with different but identifiable lineage. Two purebred dogs are sometimes deliberately crossed in hopes of creating a new breed such as the cockapoo, one part cocker spaniel, one part poodle. But despite what those who breed and sell crossbreeds might try and tell you, these are not, nor will they likely ever be, recognized as purebred dogs.

Typically found in animal shelters, mixed-breed dogs are usually great companions that are often in need of good homes. They are less expensive, generally healthier, and perhaps even brighter than some purebred dogs. Although they're not bred for a specific purpose, mixed breeds such as this puppy can make excellent family pets.

for dogs. Performed under anesthesia when the dog is about four months of age, the ears are cut, splinted upright, and taped. In guard dogs such as Dobermans and boxers, and even the gentle Great Dane, alert-looking ears supposedly help create a more ferocious appearance. Although many people don't realize it, a Doberman's ears are naturally big and floppy. Tails are usually docked when the puppy is a few days old. A vet cuts off the tail at one of the coccygeal vertebrae without the anaesthesia that could be dangerous to a dog so young.

Apart from injury prevention, defenders of these practices, including the AKC, argue that these age-old procedures help define the appearance and character of certain breeds. For now, the controversy rages. Unless kennel clubs change their standards, which seems unlikely at the moment, cropping and docking will continue. If there is one thing wanted by breeders and owners of purebreds, it's for their dogs to meet the standard. Serious dog show exhibitors, in particular, know that they must have their puppy's tail docked and ears cropped if its breed standard calls for that. In shows sponsored by the kennel clubs upholding these traditions, failure to comply might penalize the dog, as the overall "look" may be different from what is customary.

TOP
·DOGS·

From the earliest days of the dog fancy, as an aid to record keeping and as a basis for comparing breeds, numerous attempts have been made to classify the various types of purebred dogs into groups. The ancient Romans classified breeds according to the original tasks or purposes for which they were best suited. They assigned dogs to one of three groups: sporting dogs, shepherd dogs, and house dogs; then they subdivided the sporting group into war dogs, scent hunters, and sight hunters. British scholar and royal physician Dr. John Caius refined the Roman system in his 1570 *Treatise Of Englishe Dogges,* the first dog book ever published. He subdivided what he called "hunters" into bloodhounds, gazehounds, greyhounds, harriers, setters, spaniels, water spaniels, and terriers and "companions" into gentle spaniels and comforters. Under "curs," Caius listed mastiffs, shepherds, and "bandogges," known today as bulldogs.

The classifying of dogs is still, for the most part, based on original purpose. However, no universal system is yet in place, and some fairly significant differences exist among the world's major registries. Featured below are descriptions of the seven breed groups used by the American Kennel Club (AKC), the world's largest national breed registry, rivaled in prestige and power only by Britain's Kennel Club. An eighth AKC group, miscellaneous, contains breeds under consideration for official recognition *(page 203)*. A look at the AKC's fifty most popular breeds follows the breed groupings, beginning with in-depth profiles of the top ten breeds. Breeds eleven through fifty are discussed in alphabetical order using charts.

SPORTING DOGS

The sporting group comprises some of the most popular breeds, including pointers, setters, retrievers, and the spaniels. Both routinely in the top five breeds, the Labrador retriever and the golden retriever together account for nearly one-quarter of the more than one million dogs registered with the AKC every year.

Alert, active, and intelligent, sporting dogs have historically been used by hunters to locate, flush, or retrieve game from land or water. Many sporting dogs are still used as hunting companions today, although their gentle natures and high level of trainability have also earned them the reputation of being among the best family dogs. These same traits often lead to some of these animals being recruited into service positions, either as helper dogs

This golden retriever demonstrates what he was bred to do. In this case, a plastic bumper replaces downed game. Like the cocker spaniel, the golden retriever has a coat that is considered less than ideal for its original purpose. A little on the long side, it is slow to dry after a foray into water and may become entangled while the dog is retrieving game in dense brush.

for the disabled or as bomb and drug sniffers for law enforcement agencies. Unfortunately, the sheer popularity of some of the sporting breeds, most notably cocker spaniels but also Labrador and golden retrievers, has led to a rise in health and behavioral problems. Some indiscriminate breeders have inflamed congenital problems by careless overbreeding, and have created unstable temperaments through simple neglect. Be especially careful in finding a reputable breeder of these dogs (*page 77*).

HOUND DOGS

The hounds are the original hunting dogs, many predating by far the gun-assisting hunters in the sporting group. There is a great deal of diversity, both behavioral and physical, within this group, a history of hunting assistance often being the only common bond among some of the hound breeds. In size, they range from the tall and lanky Irish wolfhound to the short-legged dachshund. A hound by origin only, the dachshund might seem to be out of place in this group, as its background as a digger accustomed to chasing foxes and badgers underground would more logically make it a terrier.

For the most part, these breeds originally assisted hunters in the field with either excellent scenting abilities or exceptional speed. Scent hounds such as bloodhounds, beagles, and foxhounds have historically aided hunters by following the scent trails left by their quarry. Today the slow, plodding bloodhound is commonly used by law enforcement to track fugitives or missing persons. Some of the oldest breeds of domestic dogs are the speedy sight hounds. Saluki and pharaoh hounds, in particular, can trace their origins back to antiquity. Images of dogs closely resembling these breeds are depicted on the walls of the tombs of the Egyptian pharaohs.

Apart from their hunting skills, most hounds make excellent pets. Reliable, sturdy, and possessing excellent stamina, they make great companions for adults and children alike. Even the famous racer, the greyhound, has proven to be a popular family pet. Those seeking one out, however, are cautioned when choosing a dog bred for the racetrack. A retired racer should be tested with small children or other pets, as some of these dogs make a habit of chasing them as they had chased mechanical rabbits in their earlier career.

WORKING DOGS

A diverse group skilled in a number of disciplines, most working breeds are robust, intelligent, and headstrong, often unsuitable for novice owners. Made up of guardians of livestock and property, police dogs, sled dogs, and rescue dogs, these workers come in all shapes and sizes, from the standard schnauzer to the Great Dane. But for the most part, these are large, powerful dogs. The Akita (Japan), the komondor (Hungary), the Portuguese water dog, the Newfoundland, the St. Bernard (Switzerland), the Alaskan Malamute, and many more make this group a veritable United Nations of dogs.

Befitting their name, a pack of English foxhounds sniffs for signs of an elusive fox on a traditional English hunt. Other hounds are also named for the prey they were first bred to pursue. Harriers were developed to hunt hares and rabbits, while otterhounds once killed otters that depleted fish stocks.

Smart, strong, and hardy northern breeds such as the Siberian husky make up part of the working dog group. Strong willed and just plain strong, these breeds can be tough to control for all but the most experienced dog owner. Female working dogs are occasionally less dominant and easier to handle.

Terriers pack a lot of energy and attitude into their rather dainty physiques. Bred to go where larger hunting hounds couldn't, dogs such as this Scottish terrier still love to dig.

Many toy dogs, including the Maltese, are thought to be of Mediterranean origin. The Romans are believed to have been among the first to breed dogs for companionship rather than for a purpose.

Without the right training, some working dogs can be difficult to handle, even dangerous. Very bright and rather determined breeds such as Rottweilers, Dobermans, and Akitas have become extremely popular, even trendy. Motivated by rising inner-city crime rates and by the rather twisted notion that owning a powerful dog somehow enhances one's social standing, all too many people have invested a lot of money in acquiring one of these dogs, but none of the time needed to learn how to control them. Only early socialization and a firm, experienced leader can convince some of these dogs that not all strangers, especially children, pose a threat.

TERRIERS

Feisty is the word most often used to describe terriers. From the latin *terra*, for earth, most terriers were originally bred to "go to ground" after burrowing vermin, larger rodents, and even foxes. These fiery little dynamos would dig up underground dens and burrows while barking furiously, forcing the inhabitants out where hunters awaited. Some breeds were even bred to finish the job themselves. Let loose in your backyard, a terrier can build an entire golf course in a day—the eighteen holes at least.

Too large to go to ground, the popular Airedale terrier puts its strength and stubborn streak to use as a surprisingly ferocious watchdog. Like most terriers, this "king of terriers" has little time for other dogs, and if not properly supervised may engage in some street brawling. If it weren't for the fact that most terriers, such as the cairn and the Norfolk, are fairly small, their tenacious nature and boundless energy would make them tough to control.

Due to some unscrupulous breeders and unmindful owners, a few breeds within the terrier group have developed rather notorious reputations. The crossing of bulldogs and terriers for the express purpose of creating fighting dogs has produced several dog breeds that can be dangerous in the wrong hands. Combining the taut muscles and compact power of the bulldog with the tenacity and aggressiveness of the terrier, some controversial bull terrier breeds have been involved in some highly publicized biting incidents, several involving small children. When these dogs bite, they don't let go. Unfortunately, these incidents tarnish the reputations of what can be friendly, stable, even calm pets. But without the right training and socialization, and in irresponsible hands, these can be dangerous dogs.

TOY DOGS

Luckily for the toy breeds, providing companionship for humans has counted as suitable employment through the ages. This has ensured the survival of the breeds without practical skills, such as the Chihuahua, pug, and Pomeranian. Many toy breeds, such as the miniature pinscher, the toy poodle, and the English toy terrier appear to be miniaturized versions of larger breeds. Ranging between under six pounds (3 kilograms) in the tiniest Chihuahua and twenty pounds (9 kilograms) in the stockiest of pugs, these diminutive dogs have made for loving companions since they were first bred centuries ago. Later some toy breeds were the lapdogs of European royalty. Today their stature makes them excellent pets for people without a lot of extra space in their homes. And despite their tendency toward yappiness, they are considered the best dogs for novice owners, though their fragility can make them less than ideal pets for families with small children.

Toy dogs' love of attention serves them well outside of their loving homes, too. Loyal and intelligent, they are great at learning tricks, and many excel in obedience competitions.

NON-SPORTING DOGS

This is the catchall group for breeds that didn't seem to fit in elsewhere, from the cuddly bichon frise, a little too big to be considered a toy, to the striking Dalmatian and the stunning but difficult Chow Chow. Their individual skills, original purposes, and temperaments are almost as varied as their origins. The poodle is by far the most popular of the non-sporting breeds. A pampered, yet surprisingly active companion today, it once was a skilled truffle hunter. In more modern times, the poodle's intelligence and trainability saw it employed in show business, commonly in circuses.

The poodle's opposite may be the bulldog. The national symbol of England, known for its strength and determination, it has been out of work since bull-baiting went out of fashion in the late nineteenth century. It now serves only as a loving, albeit somewhat sedentary, companion.

HERDING DOGS

Nonexistent in European registries, where most of these breeds are classified as working dogs, this group was created in 1983 by the AKC to honor what is one of the oldest of dog professions: herding livestock. The sheepdogs and cattle dogs of the world are here, including the much-loved collie breeds and those royal favorites, the corgis. Unlike the livestock guardians that simply stand sentry, herding dogs actively round up cattle and sheep with frantic running, eye contact, and aggressive barking.

Some of the more intelligent dog breeds belong to this group, including the popular German shepherd dog, perhaps most famous for its police work, and what is arguably the most intelligent of all breeds, the Border collie. Although most of these are now simple companion dogs that have never even seen a sheep, the instinct to herd in some of them can be strong. Where no livestock exists, children and adults alike may be rounded up into corners or even tight circles by these serious, tireless workers. They require owners who are skilled at training and willing to give them "work" that rewards their instincts.

Association of the Dalmatian, a non-sporting group dog, with firemen and firehouses, in particular, predates the motor car. Before that time, they worked as carriage dogs, running ahead to clear the road for horse-drawn carriages, including fire wagons. Following that tradition, many fire stations today keep Dalmatians as mascots.

Barking furiously, this bearded collie rounds up a flock of sheep. Bred to assist farmers and shepherds, these highly intelligent dogs have also done duty on police forces, rescue teams, and television and movie acting crews.

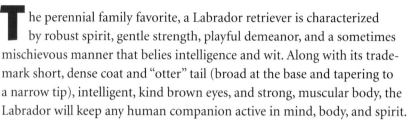

LABRADOR
·RETRIEVER·

Labrador retriever

The perennial family favorite, a Labrador retriever is characterized by robust spirit, gentle strength, playful demeanor, and a sometimes mischievous manner that belies intelligence and wit. Along with its trademark short, dense coat and "otter" tail (broad at the base and tapering to a narrow tip), intelligent, kind brown eyes, and strong, muscular body, the Labrador will keep any human companion active in mind, body, and spirit.

Labs originated in the early nineteenth century in Newfoundland, Canada, as descendants of dogs taken and found there by explorers, fisherman, and hunters. Canadian salt-cod traders brought Labs over to Dorset, England, where they were prized for their exceptional retrieving, running, swimming, and fighting abilities. Strict British quarantine laws that forbade the dogs to be brought home, along with a heavy dog tax, signaled the end of Labs in Newfoundland. There was no attempt to keep the remaining Lab line pure, but the dog's characteristics predominated in most interbreeding attempts. Eventually dog fanciers drew up a standard to discourage further cross-mixing of this rambunctious breed. Bred as gun dogs in England, many British Labs were brought to the U.S. in the 1920s and 1930s. The British Kennel Club recognized the Labrador as a distinct breed in 1903, the American Kennel Club (AKC) in 1917.

The Lab is prized for its stunning athletic physique, its graceful gait, and its agility. A sturdy barrel chest, tapering into a slim abdomen, and strong, muscular neck and legs are the hallmarks of this medium-sized hunting breed. The black Lab may have been noticed first, but its yellow and chocolate-coated cousins have since gained equal popularity.

Labs love rigorous play and exercise, and their calm social nature makes them excellent companions. A Lab is generally eager to please, unaggressive, and easily trained, although as adolescents they can be dominant and very active, posing problems for inexperienced owners. Your favorite rug, shoes, or books may be casualties of the war on boredom, so give attention, affection, and outlets for the dog's playful energies. The Lab's strength, intelligence, compassion, dependability, and even temperament make the breed a prime choice for guide dogs for the blind, narcotics detection, and arduous search and rescue work.

The sturdy Lab is susceptible to several hereditary diseases, among them hip and elbow dysplasia, bloat, epilepsy, and eye problems such as cataracts. In addition, Labs are prone to obesity, especially as they age.

GOLDEN
·RETRIEVER·

Versatile, talented, friendly, and beautiful, the golden retriever is probably one of the most photographed breeds among all those recognized by the AKC. Leaf through any magazine and chances are, the sunny "golden dog" will more often than not be looking back at you.

The first litter of yellow retriever pups was bred in 1886 by avid hunter and waterfowler Lord Tweedmouth, who crossed a flat-coated retriever and a liver-colored tweed water spaniel. He then line-bred on any yellow pups produced by one of the original females and her littermates. His goal was to breed a yellow dog that could retrieve game from both the field and the rough English seacoast. This popular "yellow retriever" sporting dog traveled to America with hunters, and received official AKC status in 1925.

With a thick coat—varying in texture from flat to wavy and in color from cream to gold—a dense, water-repellent undercoat, tufts of hair between the pads of its feet, good feathering (longer hairs) on the forelimbs and tail, and muscular limbs and neck, this medium-sized dog was born to retrieve in all weather conditions. A powerful muzzle allows it to hold onto whatever it fetches, but a gentle, sturdy mouth keeps even the most delicate of items from being crushed between mighty jaws.

Goldens make great family dogs. They are eager to please, intelligent, playful, and very trainable. You can trust them not to snap or bite, even with the rowdiest of children. Although your golden will demand your affection and attention, he will return it tenfold. However, if he is bored, he may turn your shoes, socks, or even a carelessly discarded sweater into a shredded mess. His retrieving instinct is strong, so be prepared to receive gifts ranging from sticks to fallen leaves during walks. In addition, you'll likely have to hold on tight to this waterfowler's leash or be dragged for an impromptu swim if your journey takes you near a lake or pool.

The golden retriever's sociable, sturdy, and intelligent nature translates into consistently high standings in all athletic and show competitions. In addition, a compassion and affinity for work allows the breed to be used as guide dogs for the blind and helpers for the disabled.

As with most widely popular, extensively bred dogs, goldens can suffer from certain inherited diseases, including hip dysplasia, skin allergies, heart and eye trouble, and von Willebrand's disease, a blood clotting disorder.

THE GOLDEN RETRIEVER AT A GLANCE

Background: Sporting group. Developed in England as a hunting dog.

Life expectancy: 10 to 15 years.

Coat type: Straight or wavy, dense outercoat with water-repellent undercoat. Moderate shedder.

Grooming requirements: Brush 3 times weekly.

Size: Males ideally 23 to 24 inches (58-61 cm) high at shoulders, and weighing 65 to 75 pounds (29-34 kg); females 21½ to 22½ inches (55-57 cm), weighing 60 to 70 pounds (27-32 kg).

Exercise needs: Thrives on four daily exercise periods totaling 2 hours. Adapts well to all environments.

Temperament: Moderate energy level. Friendly, outgoing, and sweet; very trainable and excels in obedience trials. Great with other animals and children. Not a watchdog.

Golden retriever

GERMAN
·SHEPHERD·

THE GERMAN SHEPHERD DOG AT A GLANCE

Background: Herding group. Originated in Germany for herding sheep.

Life expectancy: 12 to 14 years.

Coat type: Weatherproof double coat, with thick undercoat and dense outercoat. High shedder.

Grooming requirements: Brush daily.

Size: Males ideally 24 to 26 inches (61-66 cm) high at shoulders, weighing 65 to 90 pounds (29-41 kg); females 22 to 24 inches (56-61 cm), weighing 55 to 80 pounds (25-36 kg).

Exercise needs: Does best with four daily exercise periods totaling 2 hours; thrives in suburbs or country.

Temperament: Very high energy level. Ultimate versatile working dog; highly intelligent; fearless and self-confident. Very willing to learn and please owner; extremely bonded to family; reserved, but not hostile with strangers. Good with children and other animals. Excellent watchdog.

The German shepherd dog first caught the public's adoring eye in the 1930s and '40s with the popular *Rin-Tin-Tin* films. Always ready for an adventure, this breed can also be gentle and affectionate. Above all, it is very smart, and will test you if you don't supply firm guidelines and training.

Known in Germany as the Deutsche shäferhund and in most of the British Commonwealth as the Alsatian, the German shepherd was the result of German cavalry officer Max von Stephanitz's efforts to produce a sheep-herding dog that was handsome, responsible, and intelligent. He succeeded by combining long-, short-, and wirehaired shepherd dogs. The breed was common throughout Germany by the time of World War I; from then on its popularity quickly spread to many other parts of the world.

Dense medium-length coats of black, sable, or black with tan or gray markings are an AKC standard of the breed, although long-haired coats and those of yellow, cream, and white are all produced as well. This noble specimen of strength and agility sports dark, intelligent, almond-shaped eyes, a lean head with long muzzle, and alert, erect ears. Along with a deep chest, well-muscled thighs, and a powerful neck, a good German shepherd is robust but never unwieldy. In fact, its easy, effortless trot is almost poetic.

Bright, active, confident, and good-natured, these dogs are suited for families that will maintain a firm, fair, and consistent training and obedience schedule. Once you win their friendship, they are loyal to a fault. They love play and strenuous exercise, and are not above destroying a shoe or two if left alone for too long. These dogs take well to people if they're socialized while they're still young, but must always be reminded who's boss. Keen senses make them well-suited to police and military work, avalanche search and rescue, and being companion guides for the blind.

While responsible breeders strive to protect the breed's fine qualities, unscrupulous and careless breeders have allowed the shepherd's inherited health problems to proliferate over successive generations. Common health concerns to watch out for include hip and elbow dysplasia, skin and eye problems, spinal-cord paralysis, epilepsy, bloat, and heart defects. In addition, poorly bred German shepherds can be overly nervous, fearful, timid, or aggressive with other dogs.

German shepherd dog

·ROTTWEILER·

Bold, intimidating, and self-assured, the Rottweiler has a sometimes unfounded reputation for violence. While it's true that the breed is widely used in protection and attack, and is even celebrated for its aggressive traits in the German Schutzhund sport that combines tracking, obedience, and protection, the short-coated "Rottie" can be a loving companion. It all comes down to breeding, early socialization, and firm training.

The breed's mastifflike precursors were used for driving and guarding cattle alongside the European-invading Roman army as early as A.D. 74. The dogs were left behind when the Romans were ousted. Bred with sheepdogs, the breed's name came about in the 1800s in the German village of Rottweil, itself named for the red tiles that were unearthed when old Roman villas were excavated. Rotties drove the butchers' cattle to and from the market, earning the nickname of "butchers' dog of Rottweil." If dog fanciers hadn't taken to this breed, it might have died out when railroads and better roads rendered their roles obsolete. Rotties became popular in Europe, notably in Germany, then were introduced to North America early in the twentieth century.

With a compact body, muscular neck and hind legs, powerful jaws, and piercing brown eyes, the Rottweiler can often appear menacing. Coarse-coated with a soft undercoat, the standard Rottie is black with distinct rust markings over the eyes and on the muzzle, chest, throat, legs, feet, and under its docked tail. Its powerful presence disguises a surprising agility. Calm, self-assured, intelligent, affectionate, loyal, and sometimes aloof, the Rottweiler needs plenty of mental and physical exercise. It is a natural protector, may be standoffish with strangers, and approaches new situations and people slowly and deliberately. Although breeders are attempting to eliminate some of its naturally aggressive tendencies, the Rottie was originally intended as a watchdog, fighter, and attacker, so a breeder must be chosen carefully. Early socialization, firm training, and obedience from a young age and continual discipline and companionship will keep your dog well behaved and satisfied. Be careful with young children, however, since sudden movements may incite his natural tendencies. This breed is used as a police and military dog, a narcotics sniffer, and a watchdog.

Rottweilers are particularly susceptible to hip dysplasia, eye problems, bloat, cancer, and disorders of the nervous system, such as spinal-cord paralysis.

THE ROTTWEILER AT A GLANCE

Background: Working group. Originally bred in Germany for guarding.

Life expectancy: 10 to 12 years.

Coat type: Short, coarse, flat outercoat; short undercoat. Moderate shedder.

Grooming requirements: Brush 3 times a week.

Size: Males ideally 24 to 27 inches (61-69 cm) at shoulder, females 22 to 25 inches (56-64 cm). Weighs between 80 and 125 pounds (36-57 kg).

Exercise needs: Does best with four periods of exercise for a total of 2 hours each day; thrives in suburbs or country.

Temperament: High energy level. Bold, determined, and confident; fiercely loyal and protective of family. Strong-willed; needs firm training. May be aggressive with other dogs; good with children if raised with them.

Rottweiler

·DACHSHUND·

THE DACHSHUND AT A GLANCE

Background: Hound group. Originated in Germany for hunting badgers.

Life expectancy: Up to 16 years.

Coat type: Three types: short and dense; long, silky hair with fine undercoat; hard, wiry outercoat with fine undercoat. Moderate shedder.

Grooming requirements: Varies from daily to weekly brushing. Professional grooming required for wirehair and longhair.

Size: Two sizes: miniature, up to 11 pounds (5 kg) at one year of age; standard, from 16 to 32 pounds (7-15 kg).

Exercise needs: Three walks totaling 45 minutes per day. Adjusts well to any living conditions.

Temperament: Moderate to high energy level. Clever, lively, courageous, bold; independent and stubborn. Good with other pets and older, considerate children; reserved with strangers. Quite barky. Excellent watchdog.

Considering its immense popularity and venerable past, you would think that the dachshund would get a little more respect. Instead its name is constantly mispronounced (properly "dacks-hoont," not "dash-hound") and it is mockingly called the sausage or wiener dog. The dachshund is probably even improperly classified. This smallest of all hounds is more akin to the terriers.

Although it currently toils mostly as a loyal, playful, clever, and good-humored family friend, the dachshund is no joke. Strong, willful, and courageous to the point of rashness, this accomplished hunter is all dog. Developed in Germany about three hundred years ago, the dachshund's native name boldly states the original purpose for which it was bred: "Dachs" for badger, "hund" meaning dog. This short-legged braveheart was actually designed to go underground in pursuit of the ferocious badger; at thirty-five to forty pounds (16 to 18 kilograms), a formidable opponent.

The first dachshunds were much larger dogs than their modern descendants. To pursue badgers into tunnels, they needed the long, low-slung bodies and short legs that make them perhaps the most unique looking of all breeds. Today's standard dachshund is built more like a reduced twenty-pound (9-kilogram) version of the original, bred to tackle slightly smaller prey. An even tinier dog, the predecessor of today's miniature dachshund, was later developed for hunting hare. Today these two sizes are available in three hair types; smooth, longhair, and wirehair. Despite some temperamental differences among the three hair varieties, all make charming and intelligent companions, with their badger-chasing ancestor's vigor, endurance, spunk, and courage buried just below the surface. There is a lot of variety in this breed's coat colors, including black, tan, and red, as well as the dapple (mottled) and brindle (striped) patterns.

Try to make training sessions especially entertaining for dachshunds; they tend to get bored very quickly. Muscular and determined, they may even try to take over the exercise. Normally hardy and healthy dogs, the long body makes some dachshunds susceptible to back problems. Without over-pampering them, try to discourage jumping from heights and keep an eye on their weight. Watch out for health problems such as hereditary epilepsy and the deafness linked to the dapple colors.

Longhair dachshund

· BEAGLE ·

A pack animal through and through, the beagle thrives on companionship, be it canine or human. If not for its tendency to wander off, following its powerful nose, and a stubborn streak that can make training a bit of a chore, the most popular of the pack hounds would give the retrievers a run for their money as the best family dogs. Robust and playful, yet exceedingly friendly and gentle with adults and children alike, the beagle is at its best in a group. Left on its own, however, this quintessential social animal can be both destructive and noisy. Vocal even at the best of times, its lonely baying call can become a neighborhood annoyance. If you are away at work all day and nobody is around to keep your dog company, don't get a beagle.

Pack hounds of the beagle type date back to medieval Europe. By the thirteenth century, packs of hounds commonly accompanied English hunters on foot or on horseback to track all manner of quarry. At the time, these more-or-less generic hounds, the probable predecessors of foxhounds, harriers, and beagles alike, were divided according to size. The larger hounds were used to hunt deer and other large game, while the smaller variety pursued hares, rabbits, and sometimes pheasant. Called "beags," the Celtic word for small, those early small hounds are thought to have borne only a slight resemblance to today's beagle. Its current small to medium well-muscled physique (no longer than it is tall), extra keen scenting ability, and joyful personality are likely the result of a more sophisticated breeding effort in the early 1800s. Today beagles are found in two sizes, under thirteen inches and thirteen to fifteen inches (33 to 38 centimeters) tall. The coat is white with patches of tan or black (or both).

Although some still hunt, individually or in the traditional pack, most beagles today are family pets. Many enjoy displaying their tracking skills in field trials. The field beagle may be harder to train and control than the more compliant show variety, making for a more difficult pet. Commands to heel may fall on deaf ears when there are rabbits, squirrels, or even cats to pursue. Their instinct to chase, and rather blindly at that, can lead to dangers, the most obvious of which is getting hit by a car. Proper fencing is crucial to keeping them safely in your yard.

The beagle's floppy ears require regular cleaning to avoid infections. When shopping around, ask about eye problems such as cataracts and glaucoma and other diseases such as hypothyroidism and epilepsy to which beagles may be genetically predisposed.

THE BEAGLE AT A GLANCE

Background: Hound group. Originated in England as a hunting dog.

Life expectancy: 12 to 15 years.

Coat type: Short, dense. Medium shedder.

Grooming requirements: Brush weekly.

Size: Two sizes: 10 to 13 inches (25-33 cm) at shoulders, weighing 18 to 20 pounds (8-9 kg); 13 to 15 inches (33-38 cm), weighing 20 to 30 pounds (9-14 kg).

Exercise needs: Four exercise periods totaling 80 minutes per day, including the occasional run. Can adjust to both city and country living.

Temperament: Moderate energy level. Good-natured, cheerful, gentle, and eager to please; loves everyone. Independent, inquisitive, curious; loves to roam and follow scent. Excellent with other pets and children. Average watchdog.

Beagle

·POODLE·

Standard poodle

Watching it prance around the show ring, its curly coat elaborately clipped and its regal head held high, it's hard to believe that the pampered poodle ever worked a day in its life. But the dandiest of all dogs was first bred as a water retriever, fetching downed game from the icy swamps and ponds of Europe—surprising considering the typical association of this breed with the idle upper class or royalty. The unusual clipping of the fur originated because of this vocation: hair was cut from certain areas to facilitate swimming and left to protect the joints and vital organs.

The poodle's origins are as murky as the waters from which it once pulled fowl. Both Germany, where its native name *pudel* means "to splash in water," and France, where its name *caniche* is derived from "duck dog," lay claim to it. The French barbet, the Hungarian water hound, the Portuguese water dog, the Irish water dog, the spaniels, and even the Maltese are among its many hypothesized ancestors. Whether it is native to France or not, the poodle was present there at least five hundred years ago in its original standard size. Around the same time, it was bred smaller to create the miniature poodle, then again for the toy poodle, a favorite of French royalty and aristocrats for centuries. Over the years, the poodle's intelligence and trainability have seen it toil as a gun dog, a prized truffle-hunter, and even as a performer. Poodle acts were a common attraction in circuses around the world in the late 1800s. Today it is primarily a companion and show dog. All three poodle sizes make up one breed. Apart from size, there is no physical difference among them. The toy's group is obvious; the standard and miniature are in the non-sporting group. Coat colors range from white, apricot, and brown to gray and black.

There is more than meets the eye when it comes to the popular poodle. Underneath its long, curly, dense, and harsh-textured coat and springy, sophisticated gait lies a surprisingly robust, alert, and intelligent dog. A noted problem solver, the poodle regularly figures among the brightest breeds in trainability tests. Despite its dainty and even slightly arrogant appearance, it's considered one of the best of companions. Playful and loyal, poodles are particularly prized by their owners for their eagerness to please. Overbreeding has produced a great many overly yappy and generally neurotic dogs. Ask to meet and observe a puppy's parents before purchasing it. Since the poodle is also subject to a number of hereditary diseases, such as epilepsy and progressive retinal atrophy, buy from only a reputable breeder.

·CHIHUAHUA·

Although three-thousand-year-old archeological evidence suggests possible Egyptian origins for the Chihuahua, the tiniest of the AKC dogs has been living near Mexico City for more than a thousand years, good enough for landed immigrant status at the very least. Discovered in the 1850s by American tourists visiting the Mexican state for which it was named, the modern Chihuahua is descended from a dog called the techichi, bred and kept by the ancient Toltecs around the tenth century in what is now Mexico. Larger than the Chihuahua and thought to be mute, the techichi was likely crossed with a small Oriental hairless breed by the Toltecs' conquerors, the Aztecs. The result was the tiny and extremely vocal dog loved around the world today.

Tipping the scales somewhere between two and six pounds (1 to 3 kilograms) and rarely reaching more than eight inches (20 centimeters) in height, the feisty Chihuahua packs a garrulous, almost terrier-like attitude. Fine-boned yet surprisingly muscular and compact, this breed is loyal, curious, and intelligent. The Chihuahua makes a great pet—but only for the right person and situation. Sometimes temperamental and even nasty when provoked, this saucy little dynamo's stature makes it the ideal apartment dog, popular with elderly couples and people living alone. Chihuahuas can sometimes take a dislike to other dog breeds. It is vitally important to socialize them with other dogs—as well as people outside the family—so they will be comfortable in their presence. Isolation can cause fear and aggression.

Mischievous and fun-loving at the best of times, the Chihuahua can also be a bossy master manipulator. Be firm and vigilant in early training to avoid living under the rule of a tiny Latin dictator in your own home. A Chihuahua will shiver when excited or nervous, but also to stay warm. Like all toy dogs, it has a high metabolism and sheds body heat rapidly. In cold weather, be sure to keep it warm with a sweater or blanket, or in a purse or pocket. Either shorthair or longhair, this breed's coat can be any solid color, and may have white markings or patches of another color.

Because of their tiny and somewhat fragile bones, Chihuahuas should not jump from high places; their limbs can break easily. The Chihuahua's current popularity risks encouraging overbreeding on the part of profit-motivated breeders. Deal only with those who regularly test their stock for genetic problems, such as heart disease, glaucoma, and luxating patella (kneecaps that dislocate), and who breed good-tempered dogs.

THE CHIHUAHUA AT A GLANCE

Background: Toy group. Originated in Mexico as companion.

Life expectancy: 12 to 15 years.

Coat type: Two types: long and short. Single or double coat. Moderate shedder.

Grooming requirements: Brush weekly.

Size: Up to 8 inches (20 cm) tall; weight not to exceed 6 pounds (3 kg).

Exercise needs: Three periods of exercise totaling 45 minutes per day. Not fond of outdoors; ideal apartment dog.

Temperament: Moderate energy level. Alert, curious, lively, and highly bonded to owner; loves to snuggle. Sometimes very noisy. Can be hard to house-train. Must be socialized at an early age. Quite fragile; best with older children. Must be protected from cold. Excellent watchdog.

Chihuahua

YORKSHIRE
·TERRIER·

Yorkshire terrier (show coat)

Now one of the tiniest and most pampered of all breeds, the Yorkshire terrier actually once made its living chasing and catching the rats that infested the coal mines of its native northern England. Despite its rather elegant appearance and a reputation as the precocious companion of the upper crust, the Yorkie today still displays all of the fire and fury that kept rodents on the run more than a century ago.

Thought to be a descendant of the small and long-coated waterside terrier, the Yorkie was created by crossing black and tan English terriers with the now extinct Clydesdale terrier (and possibly the Skye and Paisley terriers) in the mid-nineteenth century. Later crosses with the Maltese are thought to have reduced its size by more than half and produced its now famous long coat. By the 1880s, the Yorkie had moved from the mines and into the laps of the rich. No longer an industrial-age ratter, it became the dainty and fashionable companion of Victorian-era English ladies.

Weighing as little as three pounds (1 kilogram), the Yorkie is the smallest of the terriers, but judging from its feisty attitude, no one has yet bothered to tell it so. This is one of the big dogs trapped in a tiny body. A compact dog, audacious and energetic, it is described as the ideal apartment dog, but also likes short vigorous walks to blow off some of its notorious steam. Alert, intelligent, and eager to play, the boisterous Yorkie responds well to training and is generally good-tempered. If provoked, though, especially by undisciplined children, it is by no means averse to responding with a sharp nip.

When it comes to coat color, the Yorkshire terrier is available in one model only. From the tip of its tail to the top of the skull, the hair is a steely blue, and its undersides are tan. Puppies are born black and tan, but change color around one year of age. Show dogs are kept with long and straight combed hair. Pet Yorkies are normally clipped to a more manageable shaggy length, but need the extra protection of a sweater in cold weather. The Yorkie's nails are a rather unique black, its alert-looking eyes are dark, and its ears are erect and V-shaped.

Unfortunately, a legacy of prolonged miniaturization has left the Yorkie victim to a number of medical problems. Gum disease and windpipe abnormalities plague these adorable dogs. Like many other pedigreed dogs, they suffer from eye problems such as cataracts and occasionally from joint problems, specifically luxating patellas (dislocated kneecaps). Seek out a breeder who regularly tests his stock for genetic defects.

·POMERANIAN·

Years of selective breeding may have succeeded in miniaturizing what was once a much larger dog, but breeders left the Pomeranian's considerable character intact. Today the tiny Pom still carries a big-dog personality around in one of the smallest of toy-dog packages.

The breed is named for Pomerania, a region near the Baltic Sea in northern Germany, where it was likely developed by breeding down the size of the similar-looking German spitz dog, also a progenitor of the keeshond. The Pomeranian's long, harsh, and protective coat and plumed, curled-back tail suggest even more Nordic origins. Its earliest ancestors were larger spitz-type sled dogs from Iceland and Lapland. Weighing about thirty pounds (14 kilograms) and often sporting an all-white coat, the first Pomeranians looked like small Samoyeds. Some were even used to herd sheep. By the middle of the nineteenth century, the Poms had spread throughout much of Europe as a companion dog. The breed's fortunes changed when Queen Victoria, a noted lover of dogs, came across a miniature, fiery-red version of it in Italy and brought it back to England. Due to the Queen's enormous popularity, the tiny terror we know as today's Pomeranian became one of Britain's favored breeds. When she died in 1901, Queen Victoria's favorite dog, a Pomeranian named Turi, was said to be by her side.

Despite its diminutive stature, the Pomeranian possesses all of the hardiness and vigor of its cold-weather forebearers. Its pointed face, tiny ears, bushy tail, black nose, and dark, sometimes black-rimmed eyes even bring the equally thick-coated arctic fox to mind. Underneath its mass of hair, the Pom is a sturdy and compact dog. Its once all-white coat lost to selective breeding, it now comes in particolor (mixed) varieties and many solid colors, the most common and popular of which is a fiery orange-sable. Docile enough to be a good children's dog, its thin bones, reduced by years of miniaturization, leave it at risk from the roughhousing of children. High-spirited and assertive, but friendly and affectionate, this breed is popular with apartment dwellers and home owners alike. No threat to intruders, it will nevertheless sound a shrill alarm when visitors, invited or not, come calling.

The feisty and highly intelligent Pomeranian is fairly easy to train. But make barking control one of your first training projects since these can be yappy dogs. The Pom can suffer from the joint (especially knee) and eye problems that plague many purebred dogs. But overall, this is a healthy breed.

THE POMERANIAN AT A GLANCE

Background: Toy group. Originated in Germany as a companion and a watchdog; .

Life expectancy: Up to 15 years.

Coat type: Long, straight, harsh outer coat with soft, dense undercoat. Moderate shedder. Annually "blows coat," when shedding is more extreme.

Grooming requirements: Brush weekly.

Size: No height standard; weighs from 3 to 7 pounds (1 to 3 kg).

Exercise needs: Three periods of exercise for a total of 45 minutes per day. Ideal apartment dog.

Temperament: Moderate to high energy level when indoors. Spirited, bold, and spunky; intelligent and highly alert. Extroverted, friendly, and affectionate; big dog in little package. Does best with older children; good with other pets. Excellent watchdog.

Pomeranian

AIREDALE TERRIER

Background: Terrier group. Originated in England to hunt rodents and game.

Life expectancy: 12 to 14 years.

Coat type: Hard, dense, wiry outercoat; softer undercoat. Low shedder.

Grooming requirements: Brush 3 times a week; Needs professional trimming, clipping, or stripping 4 times a year.

Size: Males ideally 23 inches (58 cm) high at shoulder, weighing 45 to 60 pounds (20-27 kg). Females slightly smaller, weighing 40 to 50 pounds (18-23 kg).

Exercise needs: Four exercise periods totaling 80 minutes per day; does best in suburbs or country.

Temperament: High energy level; lively and fun-loving with lots of drive and stamina. Clownish in play; can be bold and stubborn. May be scrappy with other animals. Good with children if raised with them; needs early socialization as well as firm, but fair, obedience training. Alert watchdog.

Airedale terrier

Akita

AKITA

Background: Working group. Originated in Japan as guard dog and hunter.

Life expectancy: 10 to 12 years.

Coat type: Medium-length, dense, weather-resistant double coat, with straight and harsh outercoat and thick undercoat. Heavy shedder.

Grooming requirements: Easy to care for with weekly brushing.

Size: Males ideally 26 to 28 inches (66-71 cm) high at shoulder, weighing 75 to 110 pounds (34-50 kg); females ideally 24 to 26 inches (61-66 cm), weighing slightly less than males.

Exercise needs: Does best with four rugged, athletic exercise periods totaling 2 hours per day; suited for suburbs or country.

Temperament: Very high energy. Powerful, reserved, protective; should be socialized to people outside family at early age. Dignified; assertive, strong-willed, and can be aggressive with other dogs; must be obedience-trained young. For experienced owners only. Best with older children. Excellent watchdog.

ALASKAN MALAMUTE

Background: Working group. Originated in Alaska as sled and draft dog.

Life expectancy: 10 to 14 years.

Coat type: Medium-length double coat; thick, coarse guard coat, dense undercoat. Heavy shedder.

Grooming requirements: Brush 3 times per week.

Size: Males ideally 24 to 26 inches (61-66 cm) high at shoulder, weighing 85 to 105 pounds (39-48 kg); females ideally 22 to 24 inches (56-61 cm), weighing 75 to 90 pounds (34-41 kg).

Exercise needs: Thrives on four exercise periods totaling about 2 hours per day. Ideal for those who enjoy outdoors; belongs in suburbs or country.

Temperament: Moderate energy level. Independent, but affectionate and very friendly; learns quickly, but gets bored easily. Tendency to roam; must be supervised outdoors. Born to pull; must learn early how to walk properly on leash. Socialization with other animals necessary; good with older, considerate children.

Alaskan Malamute

Australian shepherd

AUSTRALIAN SHEPHERD

Background: Herding group. Developed in United States as livestock herder.

Life expectancy: 14 to 16 years.

Coat type: Weather-resistant, medium-length, straight to wavy double coat. Heavy shedder.

Grooming requirements: Needs weekly brushing.

Size: Males ideally 20 to 23 inches (51-58 cm) high at shoulder, weighing 45 to 65 pounds (20-29 kg); females ideally 18 to 21 inches (46-53 cm), weighing 35 to 55 pounds (16-25 kg).

Exercise needs: Thrives on four energetic, active exercise periods totaling 2 hours per day. High stamina; best suited to suburbs or country.

Temperament: High energy level. Very intelligent, learns very quickly, and excels in obedience training and dog sports; highly focused and single-minded when working. Good with other animals and children, but can be overprotective and quick to nip. Makes great family pet if exercise needs met. Excellent watchdog.

BASSET HOUND

Background: Hound group. Originated in France as a hunting dog.

Life expectancy: 8 to 12 years.

Coat type: Smooth, short, and dense. Medium shedder.

Grooming requirements: Weekly brushing, paying special attention to ear cleaning.

Size: Males should not exceed 14 inches (36 cm) in height at shoulder; weigh 40 to 65 pounds (18-29 kg). Females slightly shorter, weigh about 10 pounds (5 kg) less.

Exercise needs: Three periods totaling about 1 hour per day; playtime also required. Suited for city living, but must be taught not to howl.

Temperament: Moderate energy level. One of mildest mannered breeds. Good-natured and peaceful; friendly with strangers; gets along well with other animals and children. Independent and stubborn; can be difficult to house-train. Hunter at heart; tends to roam. Not a watchdog.

Basset hound

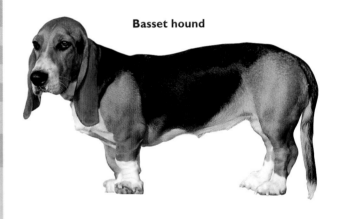

Bichon frise

BICHON FRISE

Background: Non-sporting group. Originated in Canary Islands as companion dog.

Life expectancy: 12 to 14 years.

Coat type: Soft and dense undercoat; coarser and curlier outercoat. Low shedder.

Grooming requirements: Daily brushing and combing. Tends to mat; requires professional grooming every 4 to 6 weeks.

Size: 9 to 12 inches (23-30 cm) high at shoulder, weighing 7 to 12 pounds (3-5 kg).

Exercise needs: Three periods totaling 1 hour per day; can be very active indoors. Does well in any environment.

Temperament: High energy level; lively and playful. Happy and cheerful; gets along well with other animals and strangers; good with considerate children. Not always easy to house-train, so patience is a must. Intelligent and enjoys attention. Alert watchdog.

BOSTON TERRIER

Background: Non-sporting group. Originated in United States as companion dog.

Life expectancy: 11 to 13 years.

Coat type: Short and smooth; no undercoat. Medium shedder.

Grooming requirements: Weekly brushing required.

Size: Varies for male and female among three height and weight classes: from 10 to 17 inches (25-43 cm) high at shoulder, weighing from 11 to 25 pounds (5-11 kg).

Exercise needs: Three exercise periods totaling about 1 hour per day; highly adaptable to either city or country living.

Temperament: High energy level; lively and playful. Affectionate and intelligent. Equally good with children and elderly; can be scrappy with other dogs. Mildly stubborn, but responds well to gentle obedience training. Sensitive to extremes in temperature. Excellent watchdog.

Boston terrier

Boxer

BOXER

Background: Working group. Originated in Germany as guard dog.

Life expectancy: 10 to 12 years.

Coat type: Short and smooth; no undercoat. Medium shedder.

Grooming requirements: Weekly brushing needed.

Size: Males ideally 22½ to 25 inches (57-64 cm) high at shoulder, weighing 65 to 80 pounds (29-36 kg); females 21 to 23½ inches (53-60 cm), weighing about 15 pounds (7 kg) less.

Exercise needs: Does best with four exercise periods totaling 2 hours per day. Highly adaptable to either city or country living.

Temperament: High energy level; bouncy and playful. Even-tempered; strong and stubborn. Alert, dignified and self-assured. Good with children; deliberate and wary of strangers; can be aggressive with strange dogs. Requires thorough obedience training. Makes an excellent watchdog.

BRITTANY

Background: Sporting group. Originated in France as hunting dog.

Life expectancy: 12 to 14 years.

Coat type: Medium length; dense, flat, or wavy—never curly. Medium shedder.

Grooming requirements: Needs weekly brushing.

Size: Both sexes vary from 17½ to 20½ inches (44-52 cm) high at shoulder, and weigh 30 to 40 pounds (14-18 kg).

Exercise needs: Does best with four periods of exercise totaling about 2 hours per day. Regular long walks a must. Can live in city if rigorously exercised.

Temperament: High energy level. Gentle, hardy, even-tempered, and intelligent; versatile. Eager to please and happiest if allowed to work. Good with other animals and children. Alert watchdog.

Brittany

BULLDOG

Background: Non-sporting group. Originally bred in Great Britain for bullfighting.

Life expectancy: 8 to 10 years.

Coat type: Straight, short, smooth coat. Medium shedder.

Grooming requirements: Needs weekly brushing. Skin folds around face need frequent cleaning.

Size: Both sexes generally about 12 to 15 inches (30-38 cm) high at shoulder; males weigh about 50 pounds (23 kg), females about 40 pounds (18 kg).

Exercise needs: Three periods totaling about 1 hour per day. Not suited for extreme weather conditions; ideally should live in air-conditioned home.

Temperament: Calm and peaceful. Very sweet, and affectionate; gentle. Stubborn; usually possessive of its food. Very reliable with children; good with other pets, but can be aggressive with strange dogs. Ideal for people looking for a quiet, sedate dog.

Bulldog

CAIRN TERRIER

Background: Terrier group. Originated in Scotland as vermin hunter.

Life expectancy: 12 to 14 years.

Coat type: Hard, weather-resistant outercoat; soft, dense undercoat. Low shedder.

Grooming requirements: Weekly brushing required. Needs professional clipping 4 times per year; may be hand stripped to pull out undercoat.

Size: Males about 10 inches (25 cm) high at shoulder, weighing 14 pounds (6kg); females ½ inch shorter, weighing about 1 pound (.5 kg) less.

Exercise needs: Three periods totaling about 1 hour per day. Can be quite active indoors; equally adaptable to city or country living.

Temperament: Moderate energy level. Plucky, hardy, spirited and bold; independent. Assertive, with typical terrier stubbornness; loves to dig. Good with children, but can be scrappy with other dogs. Not for those seeking a quiet, calm companion. Average watchdog.

Cairn terrier

CHESAPEAKE BAY RETRIEVER

Background: Sporting group. Developed in United States as hunting dog.

Life expectancy: 10 to 12 years.

Coat type: Water-resistant, oily, coarse, and slightly wavy outercoat covers dense woolly undercoat. Medium shedder.

Grooming requirements: Brush weekly.

Size: Males ideally 23 to 26 inches (58-66 cm) high at shoulder, weighing 65 to 80 pounds (29-36 kg); females 21 to 24 inches (53-61 cm), weighing 55 to 70 pounds (25-32 kg).

Exercise needs: Does best with three periods totaling at least 2 hours per day; almost inexhaustible. Ideal for very active family; does best in suburbs or country.

Temperament: High energy level. Most rugged and powerful of retrievers; courageous; devoted to family; willing worker; loves water. Reserved and protective with strangers; should be socialized to people at early age; can be aggressive with other dogs. Obedience training necessary as puppy. Excellent watchdog.

Chesapeake Bay retriever

CHINESE SHAR-PEI

Background: Non-sporting group. Originated in China as fighting dog.

Life expectancy: 7 to 12 years.

Coat type: Two accepted types: "horse coat," short and smooth; "brush coat," longer, but less than 1 inch in length. Both coats extremely harsh to touch. Medium shedder.

Grooming requirements: Brush weekly. Needs daily cleaning of skin folds, and monthly bathing with anti-bacterial shampoo.

Size: Males 18 to 20 inches (46-51 cm) high at shoulder, weighing 40 to 55 pounds (18-25 kg); females slightly smaller.

Exercise needs: Four periods totaling about 90 minutes per day. Suited to both city and country living.

Temperament: Moderate energy level. Confident, serious, and independent. Extremely devoted to family. Early socialization minimizes natural aggressive tendencies toward other animals. Aloof with strangers; excellent watchdog. For experienced owners only.

Chinese shar-pei

Chow Chow

CHOW CHOW

Background: Non-sporting group. Originated in China as hunter and guard dog.

Life expectancy: 8 to 15 years.

Coat type: Two types of coat: rough and smooth. Both double coated, with coarse and abundant outer layer and soft, thick, and woolly undercoat. Heavy shedder.

Grooming requirements: Brush 3 times per week.

Size: Males ideally 19 to 22 inches (48-56 cm) high at shoulder, weighing 45 to 65 pounds (20-29 kg); females 18 to 20 inches (46-51 cm), weighing 35 to 55 pounds (16-25 kg).

Exercise needs: Four periods totaling about 80 minutes per day. Adapts easily to city or country living.

Temperament: Moderate energy level. Guard dog by nature; proud; aloof with strangers; very devoted to family, but can be aggressive and stubborn. Can be unpredictable toward other animals. Better with older children. For experienced owners only.

COCKER SPANIEL

Background: Sporting group. Developed in England as hunting dog.

Life expectancy: 12 to 14 years.

Coat type: Silky, flat, or slightly wavy medium-length coat with good feathering. Moderate shedder.

Grooming requirements: Daily brushing and frequent ear cleaning. Needs professional grooming every 2 months for clipping or hand stripping.

Size: Males ideally 14½ to 15½ inches (37-39 cm) high at shoulder, weighing 25 to 30 pounds (11-14 kg); females 13½ to 14½ inches (34-37) high, weighing slightly less.

Exercise needs: Three periods totaling about 45 minutes per day. Easily adaptable to either city or country living.

Temperament: Moderate energy level. Lively, cheerful, gentle, and playful; eager to please. Most get along well with strangers and other animals and are good with children; poorly bred cockers can have aggressive traits. Early socialization and careful selection vital.

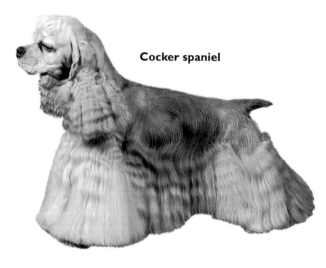

Cocker spaniel

Collie

COLLIE

Background: Herding group. Originally bred in Scotland for herding sheep.

Life expectancy: 10 to 12 years.

Coat type: Two types: smooth and rough. Smooth: short, smooth coat. Rough: double coated, with outercoat straight and harsh to touch, and undercoat thick and dense. Heavy shedder.

Grooming requirements: Smooth: easy to care for; requires weekly brushing. Rough: needs thorough brushing every day.

Size: Males ideally 24 to 26 inches high (61-66 cm) at shoulder, weighing 60 to 75 pounds (27-34 kg); females ideally 22 to 24 inches high (56-61 cm), weighing 50 to 65 pounds (23-29 kg).

Exercise needs: Four daily periods totaling about 80 minutes. Can adjust to city living, but does best in suburbs or country.

Temperament: Moderate energy level. Noble and gentle; highly intelligent and sensitive; responds well to gentle obedience training. Friendly and eager to please, but can be shy and nervous if not properly socialized. Good with considerate children; gets along well with other pets, but may try to herd smaller animals. Alert watchdog.

DALMATIAN

Background: Non-sporting group. Bred in the Balkans as versatile working dog.

Life expectancy: 12 to 14 years.

Coat type: Short and dense. Moderate shedder.

Grooming requirements: Weekly brushing.

Size: Males ideally 23 to 24 inches (58-61 cm) high at shoulder, weighing 45 to 65 pounds (20-29 kg); females ideally 22 to 23 inches (56-58 cm) high, weighing 45 to 55 pounds (20-25 kg).

Exercise needs: Does best with four periods totaling about 2 hours per day; lengthy running ideal. Thrives in suburbs or country with athletic owner.

Temperament: Very high energy level, but typically quiet and reserved; lots of stamina. Sometimes aloof with strangers. Good with other pets and loves horses, but may be aggressive with strange dogs. Good with older, considerate children. Early socialization and firm training important. Alert watchdog.

Dalmatian

Doberman pinscher

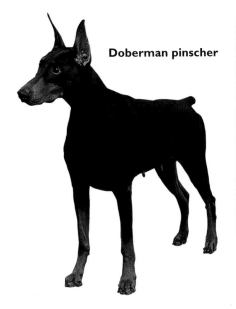

DOBERMAN PINSCHER

Background: Working group. Originated in Germany as guard dog.

Life expectancy: 12 to 15 years.

Coat type: Short, smooth coat. Moderate shedder.

Grooming requirements: Weekly brushing needed.

Size: Males ideally 26 to 28 inches (66-71 cm) high at shoulder, weighing 70 to 75 pounds (32-34 kg); females ideally 24-26 inches (61-66 cm) high, weighing 60 to 65 pounds (27-29 kg).

Exercise needs: Four periods totaling about 80 minutes per day. Can adjust well to city or country living.

Temperament: High energy level; strong, speedy, and agile. Does extremely well in obedience work. Versatile; loyal and highly intelligent. Good with children and other pets; but may be aggressive with strange dogs. Protective of family and home, making for excellent watchdog.

ENGLISH SPRINGER SPANIEL

Background: Sporting group. Bred in Great Britain for hunting.

Life expectancy: 12 to 14 years.

Coat type: Medium-length flat or wavy outercoat with feathering; short, dense undercoat. Medium shedder.

Grooming requirements: Brushing 3 times per week and professional clipping 4 times per year.

Size: Males ideally 20 inches (51 cm) high at shoulder, weighing 49 to 55 pounds (22-25 kg); females ideally 19 inches (48 cm) high, weighing slightly less.

Exercise needs: Does best with four periods totaling about 2 hours per day. Can adjust to both city and country living.

Temperament: High energy level. Friendly, affectionate, eager to please. Willing worker, quick to learn. Good with other animals and children. Somewhat reserved with strangers. Some poorly bred springers known for unpredictable aggression. Excellent watchdog.

English springer spaniel

German shorthaired pointer

GERMAN SHORTHAIRED POINTER

Background: Sporting group. Bred in Germany for hunting.

Life expectancy: 12 to 14 years.

Coat type: Short, dense coat. Medium shedder.

Grooming requirements: Weekly brushing needed.

Size: Males ideally 23 to 25 inches (58-64 cm) high at shoulder, weighing 55 to 70 pounds (25-32 kg); females ideally 21 to 23 inches (53-58 cm) high, weighing 45 to 60 pounds (20-27 kg).

Exercise needs: Does best with four periods totaling about 2 hours per day. Thrives in suburbs or country; too active for small apartment.

Temperament: High energy level; tireless. Friendly and good-natured. Can be stubborn, but highly trainable. Good with other animals and children if raised with them. Can be aloof with strangers. Barks a lot. Good watchdog.

GREAT DANE

Background: Working group. Originated in Germany to hunt wild boar.

Life expectancy: 7 to 10 years.

Coat type: Short, dense coat. Moderate shedder.

Grooming requirements: Weekly brushing.

Size: Males ideally 32 inches (81 cm) or taller at shoulder; females ideally 30 inches (76 cm) or taller. No official weight standard, but varies from 100 to 150 pounds (45-68 kg).

Exercise needs: Thrives on four periods totaling about 2 hours per day. Can adjust well to both city and country living.

Temperament: Moderate energy level. Gentle giant; sweet-natured, friendly, noble, and majestic. Tends to "lean" on favorite people; big lap dog. Needs firm but gentle handling as puppy to develop into well-mannered adult. Excellent with other pets and children. Courageous and protective; superb watchdog.

Great Dane

Great Pyrenees

GREAT PYRENEES

Background: Working group. Originated in France as flock protector.

Life expectancy: 10 to 12 years.

Coat type: Weather-resistant double coat consists of long, thick, and coarse outer-coat of straight or wavy hair and thick, dense undercoat. Heavy shedder.

Grooming requirements: Daily brushing.

Size: Males ideally 27 to 32 inches (69-81 cm) high at shoulder, weighing 100 to 125 pounds (45-57 kg); females ideally 25 to 29 inches (64-74 cm) high, weighing 85 to 115 pounds (39-52 kg).

Exercise needs: Thrives on four periods totaling about 2 hours per day; long walks are advised. Does best in suburbs or country environment.

Temperament: Moderate energy level. Confident, strong-willed, and independent. Highly territorial and protective of family pets, but will chase away others; reserved with strangers. Usually gentle and patient with children. Very trainable, but must respect owner or will not obey. Excellent watchdog.

ITALIAN GREYHOUND

Background: Toy group. Originated in Mediterranean basin as companion.

Life expectancy: 10 to 15 years.

Coat type: Soft short coat. Minimal shedder.

Grooming requirements: Weekly brushing.

Size: Desired height 13 to 15 inches (33-38 cm); weight from 8 to 15 pounds (4-7 kg).

Exercise needs: Three periods totaling about 45 minutes per day. Can adjust to any living situation.

Temperament: Calm, but alert and playful. Sensitive, mild mannered, cuddly, peaceful, and gentle. Very affectionate; craves attention of owner. Highly intelligent; learns quickly with gentle methods. May be a bit aloof around strangers; good with older considerate children and other animals.

Italian greyhound

Lhasa Apso

LHASA APSO

Background: Non-sporting group. Originated in Tibet as guard dog.

Life expectancy: 12 to 18 years.

Coat type: Long, straight, and harsh coat; double-coated. Minimal shedder.

Grooming requirements: Daily brushing. Tendency to mat; needs professional clipping 4 to 6 times a year.

Size: Variable: males, 10 to 11 inches (25-28 cm) at shoulder; females, slightly shorter. Weighs from 12 to 18 pounds (5-8 kg).

Exercise needs: Three periods totaling about 1 hour per day. Can adjust to both city and country living.

Temperament: Moderate energy level. Bold, assertive, independent, stubborn, and hardy. Not cuddly lap dog; can be quick tempered and snappy. Good with other animals and older children. Wary of strangers. Excellent watchdog.

MALTESE

Background: Toy group. Originated on island of Malta as companion.

Life expectancy: Long-lived; up to 20 years.

Coat type: Long, flat, straight, and silky single coat. Minimal shedder.

Grooming requirements: Daily brushing. Tends to mat; requires professional clipping 4 to 6 times a year.

Size: No height standard; may weigh from 4 to 7 pounds (2-3 kg).

Exercise needs: Three exercise periods totaling about 45 minutes per day. Ideal apartment dog.

Temperament: Moderate energy level. Gentle, cheerful, and extroverted; non-aggressive by nature. Not hyperactive or high-strung like some other toy breeds. Friendly with everyone, but will alert owner to presence of others. Very fragile; best with older considerate children, but adult home ideal.

Maltese

Mastiff

MASTIFF

Background: Working group. Bred in Britain as watchdog.

Life expectancy: 8 to 10 years.

Coat type: Coarse, straight outercoat; short, dense undercoat. Moderate shedder.

Grooming requirements: Weekly brushing.

Size: Males at least 30 inches (76 cm) high at shoulder; females at least 27½ inches (70 cm) high. Weight varies from 170 to 200 pounds (77-91 kg).

Exercise needs: Thrives on four exercise periods totaling about 2 hours per day. Adapts to both city and country environments.

Temperament: Moderate energy level. Easy-going, dignified, and good-natured. Intelligent. Not excitable nor overly agile. Excellent with children; reserved with strangers and manifests ancient fighting spirit with other dogs. Reliable guard dog.

MINIATURE PINSCHER

Background: Toy group. Originated in Germany as ratter and companion.

Life expectancy: About 15 years.

Coat type: Short, dense, and smooth coat. Low shedder.

Grooming requirements: Weekly brushing.

Size: Ideally 10 to 12½ inches (25-32 cm) high at shoulder; no weight standard, but typically 8 to 10 pounds (4-5 kg).

Exercise needs: Three periods totaling about 45 minutes per day; can be fulfilled in small space. Adjusts to both city and country living.

Temperament: Very high energy. Highly intelligent and alert. Headstrong, feisty, and courageous; cocky attitude. Very vocal. Good with other animals, but may challenge bigger dogs. Does best with older children. Excellent watchdog.

Miniature pinscher

Miniature schnauzer

MINIATURE SCHNAUZER

Background: Terrier group. Originally bred in Germany as ratter.

Life expectancy: 12 to 14 years.

Coat type: Double-coated: harsh wiry outercoat with softer undercoat. Low shedder.

Grooming requirements: Brushing 3 times per week with professional clipping every 2 months. May be hand stripped by groomer.

Size: Ideally 12 to 14 inches (30- 36 cm) high at shoulder; 13 to 20 pounds (6-9 kg).

Exercise needs: Four periods totaling at least 1 hour per day. Equally adaptable to city and country.

Temperament: High energy level; inquisitive, and playful. Hardy, robust, stubborn, and fearless. Alert and spunky; not mellow lap dog. Generally good with other pets, but can be territorial; good with children. Very keen watchdog.

PAPILLON

Background: Toy group. Originated in France as companion.

Life expectancy: 13 to 16 years.

Coat type: Long, fine, silky, and straight hair; no undercoat. Low shedder.

Grooming requirements: Weekly brushing.

Size: 8 to 11 inches (20-28 cm) at shoulder. Weight proportionate to height: males typically from 8 to 10 pounds (4-5 kg); females from 7 to 9 pounds (3-4 kg).

Exercise needs: Three periods totaling about 45 minutes per day. Does well in both city and country environments.

Temperament: Moderate energy level. Happy, friendly, and outgoing. Highly intelligent and easy to train; excels in obedience and agility. Good with other animals, but may challenge larger dogs. Good with older children. Gentlest and most obedient of toy breeds. Excellent watchdog.

Papillon

Pekingese

PEKINGESE

Background: Toy group. Originally bred in China as companion.

Life expectancy: 12 to 14 years.

Coat type: Long, coarse, straight outercoat with thick, softer undercoat. Moderate shedder.

Grooming requirements: Weekly brushing.

Size: Up to 9 inches (23 cm) at shoulder; both males and females should weigh no more than 14 pounds (6 kg).

Exercise needs: Three periods totaling about 45 minutes per day. Does well in any environment; perfect apartment dog.

Temperament: Calm; dignified, regal, and sedate. Assertive, independent, self-confident, and stubborn. Accepts other animals and good with older children. Can be territorial and wary of strangers. Needs protection from heat.

PEMBROKE WELSH CORGI

Background: Herding group. Originated in Wales for herding cattle.

Life expectancy: 12 to 18 years.

Coat type: Short, thick, weather-resistant undercoat with coarser, medium-length outercoat. Moderate to high shedder.

Grooming requirements: Brushing 3 times per week.

Size: 10 to 12 inches (25-30 cm) high at shoulder. Weight in proportion to size, not exceeding 30 pounds (14 kg) for males, 28 pounds (13 kg) for females.

Exercise needs: Three periods totaling about 45 minutes per day. Does well in both city and country situations.

Temperament: Moderate energy level. Affectionate, intelligent, and loyal. Hardy, spirited; very adaptable. Big-dog personality in small-dog body. Good with other animals and children. Territorial; alert watchdog.

Pembroke Welsh corgi

Pug

PUG

Background: Toy group. Originated in China as companion.

Life expectancy: 12 to 14 years.

Coat type: Short and smooth; with undercoat. Moderate shedder.

Grooming requirements: Weekly brushing and daily cleaning of facial wrinkles.

Size: 10 to 12 inches (25-30 cm); 14-18 (6-8 kg) pounds.

Exercise needs: Three periods totaling about 45 minutes per day. Does well in both city and country settings; ideal apartment dog.

Temperament: Calm; easy-going, even-tempered, comical, and friendly with all. Cuddly lap dog. Equally good with children and elderly; one of most stable of toy breeds. Not a guard dog. Requires protection from extreme heat.

ST. BERNARD

Background: Working group. Originated in Switzerland as rescue dog.

Life expectancy: 10 to 12 years.

Coat type: Two types: shorthaired (short, thick, and dense) and longhaired (medium length and slightly wavy); both have undercoats. High shedder.

Grooming requirements: Brushing 3 times per week.

Size: Males at least 27 inches (69 cm) high at shoulder; females ideally 25 inches (64 cm); weight varies from 150 to 180 pounds (68-82 kg).

Exercise needs: Thrives on four periods totaling about 2 hours per day. Does best in suburbs or country.

Temperament: Moderate energy level. Sensible, gentle, and patient. Usually fine with other animals and children. Because of enormous size, must be trained early in puppyhood. Excellent guard dog. Heavy drooler. Thrives in colder climates.

St. Bernard

Scottish terrier

SCOTTISH TERRIER

Background: Terrier group. Originated in Scotland to hunt foxes and vermin.

Life expectancy: 12 to 14 years.

Coat type: Hard wiry outercoat with soft, dense undercoat. Low shedder.

Grooming requirements: Brushing 3 times per week. Professional clipping needed every 2 to 3 months; may be hand stripped.

Size: About 10 inches (25 cm) high at shoulder for both sexes. Males weigh 19 to 22 pounds (9-10 kg), females 18 to 21 pounds (8-10 kg).

Exercise needs: Three periods totaling about 45 minutes per day. Does well in both city and country.

Temperament: Moderate energy level. Dignified, independent, and self-assured. Very alert, stubborn, and fearless. Very protective of owner; reserved with strangers. Good with older children; can be scrappy with other dogs. Likes to dig. Excellent watchdog.

SHETLAND SHEEPDOG

Background: Herding group. Originated in Scotland as sheep herder.

Life expectancy: 12 to 14 years.

Coat type: Long, straight, harsh outercoat; short, dense undercoat. High shedder.

Grooming requirements: Brushing 3 times per week.

Size: From 13 to 16 inches (33-41) high at shoulder; weight up to 20 pounds (9 kg).

Exercise needs: Four periods totaling about 80 minutes per day. Can do well in both city and country.

Temperament: Very high energy level. Highly sensitive and intelligent; learns quickly. Very willing and eager to please; one of most successful of all dogs in obedience trial competitions. Gets along well with other animals as well as considerate children; reserved, often timid with strangers. Enjoys barking. Excellent watchdog.

Shetland sheepdog

Shih Tzu

SHIH TZU

Background: Toy group. Originated in Tibet as companion dog.

Life expectancy: 12 to 14 years.

Coat type: Luxurious, dense, long, and flowing outercoat with wavy undercoat. Low shedder.

Grooming requirements: Daily brushing. Has tendency to mat; requires professional clipping 4 to 6 times per year.

Size: 8 to 11 inches (20-28 cm) high at shoulder for both sexes; weight from 9 to 16 pounds (4-7 kg).

Exercise needs: Three periods totaling about 45 minutes per day. Suited to any environment; ideal apartment dog.

Temperament: Moderate energy level. Outgoing, friendly, happy, affectionate, and trusting toward all. Lively and alert; good with other animals and children. Small, but not too fragile. Protect from extreme high temperatures. Not a guard dog.

Siberian husky

SIBERIAN HUSKY

Background: Working group. Originated in Siberia as sled dog.

Life expectancy: 10 to 12 years.

Coat type: Medium-length double coat made up of coarse guard hairs and soft, fluffy insulating undercoat. High shedder.

Grooming requirements: Brushing 3 times per week.

Size: Males ideally 21 to 23½ inches (53-60 cm) high at shoulder, weighing 45 to 60 pounds (20-27 kg); females ideally 20 to 22 inches (51-56 cm) high, weighing 35 to 50 pounds (16-23 kg).

Exercise needs: Does best with four rigorous periods totaling about 2 hours per day. Best suited for active owners; thrives in country environment.

Temperament: Very high energy level. Bred to run; inexhaustible. Gentle; friendly with everyone; very intelligent and highly independent. Easily bored; can be very destructive. Escape artist; digger; loves to roam and must be carefully supervised. Good with children and dogs, but be careful with little pets. Likes to howl and "talk." Not a watchdog.

VIZSLA

Background: Sporting group. Originally bred in Hungary for hunting.

Life expectancy: 13 to 15 years.

Coat type: Short, dense coat. Minimum shedder.

Grooming requirements: Weekly brushing.

Size: Males ideally 22 to 24 (56-61 cm) inches high at shoulder, females ideally 21 to 23 inches (53-58 cm) high. Weight varies from 45 to 65 pounds (20-29 kg).

Exercise needs: Does best with four rigorous periods totaling at least 2 hours per day; needs active owner. Best suited for suburbs or country.

Temperament: Very high energy level. Lots of stamina; agile and light-footed. Gentle, good-natured, and demonstrative with family; very dedicated and protective. Highly intelligent, but easily distracted. Gets along well with other animals and children. Alert watchdog.

Vizsla

Weimaraner

WEIMARANER

Background: Sporting group. Originally bred in Germany for hunting.

Life expectancy: 12 to 14 years.

Coat type: Short, dense coat. Moderate shedder.

Grooming requirements: Weekly brushing.

Size: Males ideally 25 to 27 inches (64-69 cm) high at shoulder; females 23 to 25 inches (58-64 cm) high. Weight varies from 50 to 75 pounds (23-34 kg).

Exercise needs: Does best with four rigorous periods totaling at least 2 hours per day. Best suited to suburbs or country.

Temperament: Very high energy level; lots of stamina. Assertive, bold, and highly intelligent, but easily distracted. Likes to roam; needs close supervision when off-leash. Good with other pets and children. Very protective of family. Excellent watchdog.

West Highland white terrier

WEST HIGHLAND WHITE TERRIER

Background: Terrier group. Originated in Scotland to hunt vermin.

Life expectancy: 14 years or more.

Coat type: Double coat; outercoat consists of straight, hard hair. Low shedder.

Grooming requirements: Brushing 3 times per week. Professional clipping required 4 to 6 times per year; may be hand stripped.

Size: Males ideally 11 inches (28 cm) high at shoulder; females ideally 10 inches (25 cm) high; can weigh up to 22 pounds (10 kg).

Exercise needs: Three periods totaling about 45 minutes per day. Good in both city and country conditions.

Temperament: High energy level. Feisty, bold, and assertive; comical and playful. Charming, cheerful, alert, courageous; hardy; not as stubborn as many terriers, but still quite independent. Good with everyone; accepting of other animals, but can be scrappy with other dogs.

POPULARITY CONTEST

Ranking dogs by their perceived popularity can be tricky business. No matter how you go about it, you are bound to raise the ire of dog lovers somewhere by excluding their favorite breed. The preceding list of the fifty most popular breeds will surely be no different. "Where on Earth," devotees of the devilishly charming star of stage and screen may ask, "is the beloved Jack Russell terrier?" And "How could the beautiful Bernese mountain dog not be in the top fifty?" Actually, the Bernese mountain dog occupied sixty-third place in the 1998 American Kennel Club (AKC) rankings, with the Jack Russell terrier a distant seventy-eighth.

Relying on statistics, specifically the number of dogs registered with a certain kennel club or registry, is perhaps the least subjective approach. But even that system has its flaws. Both geographic location and the differences among the various registries around the world will influence the rankings. The English foxhound may be quite popular in its native England, but in the United States, where traditional fox hunting is rarely practiced, very few are registered with the AKC.

Exactly which stats are applied can change the rankings a great deal. In this book, the venerable American Kennel Club's list of top fifty breeds, according to the number of dogs registered there in 1998, was used. There are more than four hundred breeds currently recognized by at least one of the dozens of kennel clubs in the world, none of which officially recognizes every single one. In Britain, The Kennel Club recognizes about forty breeds more than the AKC does. Extremely popular in Europe, the Jack Russell terrier is not yet officially recognized by the AKC, a fact sure to influence the number of registrations in the United States. The feisty and intelligent little dog, developed by Reverend Jack Russell in the nineteenth century in England, currently resides in the AKC's Miscellaneous category. Before graduating to the terrier group, like any other breed seeking official recognition, it must be the subject of substantial national interest, be promoted by an active breed club, and be bred serious-

ly over a substantial geographic area. Simply being popular just doesn't cut it in the AKC.

For information on breeds outside the top fifty, the AKC publishes profiles of the more than 145 breeds it recognizes, including those in the Miscellaneous category, both in book form and on the internet. The American Rare Breed Association (see page 211) can provide information on many of the more exotic breeds.

Bernese mountain dog

DOG OWNER'S

RESOURCE

GUIDE

◆ ◆ ◆

IN CASE OF
·EMERGENCY·

To be prepared for a medical emergency, have a first-aid kit *(box, right)* handy so you can stabilize your dog until you can get him to a vet *(page 162)*. In case of an environmental emergency, such as a severe storm, flood, or hurricane, think out an evacuation plan for your dog in advance. If emergency personnel evacuate your family, they may refuse to allow you to bring your pet, and most shelters set up for evacuees are not permitted to take in animals. If there is an early warning, while you still have the option of removing your dog from the impending danger area, take him to the safety of a friend's house, or drop him off at a vet or boarding facility. Or find out if any humane societies are boarding animals for the duration of the disaster.

In the event of a quick exit, be prepared to gather the following items: the dog carrier or cage lined with an easily dried blanket (or a sleeping bag); a spare collar or harness and sturdy, long leash, preferably one that is retractable; two unbreakable bowls with at least one week's supply of food and water in closeable containers; a can opener and spoon if necessary; dog treats; and a few toys. You'll also need some plastic bags, paper towels, grooming supplies, and dish soap. Keep handy your dog's medical and vaccination records—most animal shelters will not take a pet without proof of up-to-date vaccinations. Of course, bring your dog's medications and ointments.

Make sure your dog is properly identified *(page 209)*, with a collar and ID tag, even if he is tattooed or microchipped. Car key tags make good temporary ID—include the phone number where he will be staying, your number, and a contact number outside the affected area in the event that phone lines are down. In case your dog gets lost, photos are helpful for finding him in a pound.

CANINE FIRST-AID KIT

Being prepared in case of emergency can save your dog's life. You can find most of the following items for your canine first-aid kit in drugstores, or simply add the appropriate items to your family's first-aid kit to serve both human and dog members.

- ◆ Scissors
- ◆ Tweezers
- ◆ Needlenose pliers
- ◆ Penlight flashlight
- ◆ Magnifying glass (type with light is best)
- ◆ Examination gloves
- ◆ Rectal thermometer and lubricant
- ◆ Isopropyl rubbing alcohol (70%)
- ◆ Hydrogen peroxide (3%)
- ◆ Povidone-iodine
- ◆ Antibiotic ointment (neomycin, polymixin, or bacitracin)
- ◆ Assorted sizes of sterile non-stick pads, gauze squares, and cotton balls
- ◆ Roller gauze (self-adhering), cotton roll, and elastic bandage
- ◆ Adhesive tape
- ◆ Cardboard or wood for splints
- ◆ Eyedropper and syringe (needle removed)
- ◆ Syrup of ipecac (Caution: Only to be given on instructions by vet or poison control center and only in dosage specified)
- ◆ Eye wash
- ◆ Styptic pencil for cut vein in nail
- ◆ Ice pack
- ◆ Large blanket
- ◆ Elizabethan collar (available for sale at many vet clinics) or bitter apple spray to prevent licking of injury
- ◆ Muzzle; or handkerchief, gauze strip, or rope for makeshift muzzle
- ◆ Board or towel for makeshift stretcher

HOUSEHOLD
· POISONS ·

Not all products that are poisonous to a dog are labeled as toxic. And some things that are safe for humans, such as medications and onions, can be deadly for your pet. Your dog doesn't necessarily have to eat or drink something to ingest it. Whatever his paws or body come into contact with can be swallowed when he is grooming himself. Unless you know that a product is safe, treat it as a potential poison. Store it in a tightly sealed container in a securely closed cabinet, preferably out of your dog's reach and line of sight.

Post the phone numbers of your vet, an emergency vet clinic, and an animal poison control center. If your dog shows signs of poisoning, such as trouble breathing, seizures, a rapid or slow heartbeat, drowsiness, drooling, or bleeding from the anus, mouth or nose, try to determine exactly what substance he ingested, and call for help. If you have the offending product, the package may contain vital information; have it on hand for

the call and take it with you to the vet. Keep a supply of syrup of ipecac, but do not make your dog vomit unless you are advised to do so. In some cases, regurgitation can worsen the problem; caustic substances can burn your dog's throat and mouth on the way back up.

The following list gives an idea of the sorts of dangerous products that must be kept out of the reach of your dog's curious paws, nose or tongue. Along with the basic principles of dog-proofing *(page 82)*, this list should enable you to make your house safer for your canine friend.

TOXIC PRODUCTS
- acetaminophen
- acetone
- ant/bug traps and baits
- anti-flea treatments
- antifreeze
- antihistamines
- anti-rust agents
- antiseptics
- arsenic
- aspirin (ASA)
- bath oil
- battery acid
- bleach
- boric acid

- brake fluid
- carbolic acid (phenol)
- carburetor cleaner
- chocolate (especially dark or bitter types)
- cleaning products
- crayons and pastels
- dandruff shampoo
- de-icers (to melt snow)
- deodorants
- deodorizers
- detergents
- diet pills
- disinfectants
- drain cleaner and opener
- dry-cleaning fluid
- dyes
- fertilizer
- fire-extinguisher foam
- fireworks
- fungicides
- furniture polish
- gasoline and motor oil
- glue and paste
- hair coloring
- heart pills
- herbicides
- ibuprofen
- insect and moth repellents
- insecticides/pesticides
- kerosene
- laxatives
- lead (also found in paint, ceramic, and linoleum)
- lighter fluid
- liniments

- lye
- matches
- medications
- mercury
- metal polish
- mineral spirits
- mothballs and repellents
- nail polish and remover
- onions
- pain relievers
- paint
- paint remover and thinner
- perfume
- permanent-wave lotion
- photographic developers
- pine-based cleaners
- pine-oil products
- plaster and putty
- rat/rodent poisons
- road salt
- rubbing alcohol
- rust remover
- shoe dye and polish
- sleeping pills
- snail or slug bait
- soap and shampoo
- solder
- solvents (e.g. turpentine)
- stain removers
- swimming pool products
- suntan lotion with cocoa butter
- toilet bowl cleaners
- weed killers
- windshield-washer fluid
- wood preservatives

PLANTS
·TO AVOID·

While dogs in the wild may instinctively avoid toxic plants, your dog may not be as savvy. Make the choices simple for your pet: keep the harmful plants listed below out of your house and garden. This list is not an exhaustive one, so always double check with your vet; some toxic plants may be particular to your area.

A dog's reaction to ingesting a toxic plant can be fairly mild, or the dog may become dehydrated, suffer from diarrhea, or even die. If your pet has eaten some dangerous greenery, contact your vet immediately.

TOXIC GREENERY
- algae
- almonds
- amaryllis
- apricots
- arrowhead vine
- asparagus fern
- autumn crocus
- azalea
- blackberry
- black-eyed Susan
- black nightshade
- bleeding heart
- boxwood

- bracken or brake fern
- buckeye
- buttercups
- cactus (spines)
- caladium
- calla lily
- castor beans
- ceriman
- charming dieffenbachia
- cherry
- Chinese evergreen
- chokecherry
- Christmas rose
- chrysanthemum
- cineraria
- clematis
- climbing nightshade
- cordatum
- corn plant
- cornstalk plant
- crabgrass
- crocus
- croton
- crown of thorns
- Cuban laurel
- daffodil
- devil's ivy
- dumb cane
- Easter lily
- elderberry
- elephant's ear
- emerald feather
- English holly
- eucalyptus
- fiddle-leaf fig
- foxglove

- glory lily
- gold dust dracaena
- helleborus
- hemlock
- holly berries
- hyacinth
- hydrangea
- iris
- ivy
- jack-in-the-pulpit
- Japanese show lily
- jasmine
- Jerusalem cherry
- jonquil
- kalanchoe
- laburnum
- lantana
- larkspur
- ligustrum
- lily of the valley
- marble queen
- marijuana
- mistletoe
- monkshood
- morning glory
- mushrooms
- narcissus
- nephthytis
- nettles
- nutmeg
- oleander
- onion
- oriental lily
- peace lily
- peach
- pencil cactus

- periwinkle
- philodendron
- plumosa fern
- poinsettia
- poison hemlock
- poison ivy
- poison oak
- pokeweed
- potato
- precatory beans
- primrose
- privet
- purple foxglove
- red emerald
- red princess
- rhododendron
- rhubarb
- rubber plants
- sago lily
- skunk cabbage
- spider plant
- spring bulbs
- string of pearls
- Swiss cheese plant
- taro vine
- tiger lily
- tinsel tree
- tobacco
- tomato plant
- tulip
- wandering Jew
- water hemlock
- wild black cherry
- wisteria
- yellow jasmine
- yew

CLEANING

·DOG STAINS·

Having a dog usually means dealing with the occasional accident, even after he's been house-trained. To prevent stains from setting in, a quick and thorough cleanup is necessary. Club soda is an efficient product to stock for this purpose, as detailed in the chart below. Soil is lifted to the surface by its bubbles, and its salts help prevent stains. It's also extremely important to obliterate all traces of odor, so keep a pet-odor neutralizer on hand. Sold at pet-supply stores, these products don't just mask the odor of urine or feces but eradicate it by breaking up the particular combination of odor molecules. If your dog's sensitive nose can detect any residue of the smell, he may continue to eliminate in the same spot. For a similar reason, avoid ammonia-based cleaners since ammonia smells similar to urine.

Pick up any solids with a spatula, a piece of stiff cardboard, or a paper towel, and then attack the stain. Don't dampen a dry stain with water; just apply the appropriate cleaner. Always spot-test the cleaner first on a concealed area of the carpet or fabric. If stains remain, use a commercial cleaner designed for pet stains and appropriate for the material. Follow label directions and do a spot test first. If an odor neutralizer can't dispense with the smell, you may have to get rid of the rug.

Your dog bed or pillow may have a removable, washable cover. If not, spot clean it with a non-toxic product. You can make your own pleasant smelling cleaning solution by adding half a teaspoon of dish soap or shampoo and a few drops of eucalyptus oil to a quart of warm water. Dry the bedding quickly—outside if possible, to freshen it.

Hair evading the vacuum can be removed with masking tape or sticky lint removers, a damp sponge or rubber gloves, or a stiff-bristled hairbrush.

CLEANUP ESSENTIALS

- ◆ Paper towels
- ◆ Spatula or cardboard
- ◆ Club soda
- ◆ Baking soda or salt
- ◆ Hydrogen peroxide (3%)
- ◆ Ammonia
- ◆ Pet-odor neutralizer (available at pet-supply stores)
- ◆ Commercial stain remover designed for pet stains
- ◆ Commercial carpet and upholstery cleaner

STAIN CLEANING METHODS

Stain	Cleaning method
Urine	Blot up liquid with paper towels. Pour club soda on the area, then blot it up with paper towels. Next, apply a pet-odor neutralizer, following the product's directions. If stains remain, use a commercial stain remover designed for pet stains.
Feces	Remove solids, then blot up any moisture with paper towels. Then follow cleaning procedure for urine.
Vomit	As quickly as possible, scoop up solids and apply baking soda or salt. When dry, vacuum up the rest. Follow by pouring club soda on the area, then blotting it up with paper towels. If stains remain, apply a nontoxic commercial carpet and upholstery cleaner or stain remover designed for pet stains.
Overturned plant	Vacuum up soil. For stains, let any remaining moist soil dry first, then vacuum again. Apply club soda and blot with paper towels. Follow with a nontoxic commercial carpet and upholstery cleaner or stain remover.
Other	For difficult-to-remove matter, follow cleaning procedure for urine. If stains remain, mix ½ cup of 3% hydrogen peroxide with 1 teaspoon of ammonia and apply solution to stain. Rinse with club soda to remove peroxide and ammonia residue.

MISSING
·DOGS·

The lure of the world beyond the backyard is too strong for some dogs, and they will take a tour of the neighborhood if the gate is left unlatched or if the fence is low enough to be jumped. Even a dog being walked off-leash can run out of your sight after something irresistible. Most dogs will return on their own, or, if properly identified, with someone's help. But if your dog isn't familiar with his surroundings—especially in new territory—he might have trouble finding his way back, or could become injured on the journey home, wander into a garage and become trapped, become disoriented and stray even farther away, or be taken to an animal shelter. Some dogs are even stolen—sometimes right out of their backyard.

Ensuring that your dog can be easily identified is your best chance of getting him back. Most cities have strict laws about keeping collars and ID tags on all dogs. The best type of collar is one of sturdy nylon or leather and should fit snugly, but you should always be able to get two fingers between the collar and your dog's neck.

Whether you choose plastic or engraved metal, a wide range of ID tags is available to go on the collar, along with the rabies vaccination tag and dog license. Although it may seem natural to include your dog's name on the ID tag, this may make it easier for thieves to coax him away. The tag should, instead, include information essential to your dog's safe return, such as your name and address, one or two reliable telephone numbers with answering machines, and even a line which reads, "Reward for Return." When you're traveling with your dog, buy replaceable key tags and list your name and a number where you can be reached at each leg of your tour.

If you're considering a more permanent means of identifying your dog, (since collars can come off or be removed by thieves,) you can choose tattoos or microchips. Relatively painless, a tattoo can be done by your vet while while your dog is under anesthesia. Or, your vet can implant a microchip into your dog's skin by injecting it between his shoulder blades. Many shelters, vets, and even medical labs will check for tattoos and run a scanner over unidentified dogs to check for a microchip, then contact the national registry where your dog's number is on file. Since microchips have been introduced, shelters have been able to return a significantly higher number of dogs to their owners than they'd been able to through collars alone. Of course, sometimes the tattoo can't be easily read, the scanner—if there is one at all—may not be compatible with the microchip, or your dog may be found by a person unfamiliar with these systems, who simply wants to phone the owner. Err on the side of caution: Always keep a collar on your dog if he's outside or if there's a chance he might slip out of the house.

Recent photos of your dog will further ensure that you can always identify him. Include some shots that clearly show his face, some that show his entire body, and some that focus on his identifying features. You may think you'd always be able to recognize him, but a scared or disturbed dog can take on a whole different look and fool both you and any animal shelter worker to whom you're describing him.

SNOOPY, COME HOME

Your dog may wander away from you in the park, but if he doesn't return within a few minutes, especially after you've called his name, stay in one place so he can find his way back to you. If a friend is with you, have him search the immediate area while you wait for your dog to return, then have your friend broaden his search, if necessary. If there is no sign of your dog after an hour or two, assume that he is lost or has been stolen. Your immediate priority should be to call your vet and local pound and shelters to report your dog as missing, then bring them a picture and visit daily to check for him. Post "Lost" signs (ideally with a picture and mention of a reward) at the vet clinic, shelter, pet stores, and around your area. Ask your neighbors to watch out for your dog.

GETTING
·A DOG·

Shelters, breeders, and rescue groups are usually good sources for acquiring a dog. Adopting from a shelter is inexpensive, and with the huge number of abandoned dogs available, you'd almost surely be saving a life. There are a number of ways to find an animal adoption agency in your area. To locate some of the bigger animal shelters, look in the phone book or Yellow Pages under headings such as "Humane Societies," "SPCA" (Society for the Prevention of Cruelty to Animals) or "ASPCA" ("CSPCA" in Canada), "Animal Organizations," "Animal Protection," "Animal Rescue," and "Animal Shelters." Many of these organizations also seek temporary foster homes for some of their dogs.

You can also call your municipal or county authorities to find out the location of the animal pound responsible for picking up strays in the district.

Many smaller animal shelters and private rescue groups, often run by volunteers, will advertise in the pet section of the classified ads in the local newspapers, or place posters on bulletin boards at pet stores and vet clinics. Also look for any pet stores that team up with local shelters, providing them with space to show their adoptive animals.

If you want a purebred dog, and are willing to pay a fair amount of money, look for breeders at dog shows and in dog magazines. However, an inexperienced dog shopper may find it hard to tell the good from the bad. The American Kennel Club (AKC) is affiliated with a national breed club for each of the breeds it recognizes; the breed club will refer you to a reputable breeder in your area. To track down a breed club, look on the AKC website, or contact them at 919-233-9767, or by e-mail at info@akc.org. The Canadian Kennel Club (CKC) can be reached at 1-800-250-8040 or 416-675-5511, or by e-mail at information@ckc.ca.

Purebred dogs are occasionally available for adoption at shelters and humane societies, although breed rescue groups often get to them first. Find the rescue group for the breed you're interested in through breeders or the national breed club for that type of dog.

The Internet is a great tool to help you adopt a dog, be it mixed breed or purebred. It doesn't take much surfing to find the many animal shelters, breed rescue groups, and breeders throughout North America. The sites listed here will give you a start; many have links to other relevant sites. Or, if you want to do a search of your own using a search engine, try key words such as "SPCA," or "Animal Shelter", or the name of a breed. If you don't want to cast such a wide net, narrow your search with a few terms such as "Dog" and "Shelter" and "Humane" and "Rescue," your city, and your state or province.

WEBSITES

www.akc.org/breeds. htm
Read profiles of each of the AKC's recognized breeds, and find links to the national breed clubs and breed rescue groups affiliated with each of them.

www.ckc.ca/scoop
Here are CKC breed profiles and contacts for breed clubs and rescue groups.

www.acmepet.com/civic/index.html
Find a listing of shelters and rescue groups in North America and around the globe at this address.

www.arkonline.com
This is an online magazine with links to animal care and welfare organizations.

www.dogdomain.com/humane.htm
This site lists humane societies and provides links to other animal shelters in North America and other countries.

www.newpet.com
This site offers practical advice for the pet owner, along with a free U.S. shelter-location service ("Match Making") activated by typing in your zip code.

www.petshelter.org
On this site you'll find, among other things, an online adoption center, a lost-and-found department, and a wide-ranging shelter directory.

FURTHER
·INFORMATION·

BOOKS

Alderton, David
Foxes, Wolves and Wild Dogs of the World
Blandford, London, 1998

Alderton, David
The Wolf Within
Howell Book House
New York, 1998

The American Kennel Club
The Complete Dog Book
Howell Book House
New York, 1998

Bauer, Erwin A.
Wild Dogs
Chronicle Books
San Francisco, 1994

Benjamin, Carol Lea
The Chosen Puppy
Howell Book House
New York, 1990

Benjamin, Carol Lea
Dog Training in 10 Minutes
Howell Book House
New York, 1997

Benjamin, Carol Lea
Second-Hand Dog
Howell Book House
New York, 1988

Bower, John and Caroline
The Dog Owner's Problem Solver
Reader's Digest
Pleasantville, 1998

Clutton-Brock, Juliet
Eyewitness Books Cat
Dorling Kindersley
London, 1991

Dutcher, Jim, with Richard Ballantine
The Sawtooth Pack
Rufus Publications
Bearsville, New York, 1996

Evans, Mark
Dog Doctor
Howell Book House
New York, 1996

Fogle, Bruce
The Complete Dog Training Manual
Dorling Kindersley
London, 1994

Fogle, Bruce
The Dog's Mind
Howell Book House
New York, 1990

Fogle, Bruce
The Encyclopedia of the Dog
Dorling Kindersley
London, 1995

Grady, Wayne
The Nature of Coyotes
Greystone Books
Vancouver/Toronto, 1994

Grambo, Rebecca L.
The Nature of Foxes
Greystone Books
Vancouver/Toronto, 1995

Grossman, Loyd
The Dog's Tale
BBC Books, London, 1993

Hoffman, Matthew, ed.
Dogs: The Ultimate Care Guide
Rodale Press
Emmaus, Pennsylvania, 1998

Kilcommins, Brian, and Sarah Wilson
Good Owners, Great Dogs
Warner Books
New York, 1992

Lowell, Michele
Your Purebreed Puppy: A Buyer's Guide
Owl Books
New York, 1990

Malone, John
The 125 Most Asked Questions About Dogs
Morrow
New York 1993

Marder, Ann
The Complete Dog Owner's Manual
Broadway Books
New York, 1997

McGinnis, Terri
The Well Dog Book
Random House
New York, 1994

Mech, L. David
The Way of the Wolf
Voyageur Press
Stillwater, Minnesota, 1991

Palika, Liz
Save That Dog!
Howell Book House
New York, 1997

Pinney, Chris C.
Guide to Home Pet Grooming
Barron's
Happauge, NY, 1990

Siegal, Mordecai, and Matthew Margolis
When Good Dogs Do Bad Things
Little Brown
New York, 1986

Simon, John M.
What Your Dog is Trying to Tell You
St. Martin's Griffin
New York, 1998

Spadafori, Gina
Dogs for Dummies
IDG Books
Foster City, CA, 1996

Thomas, Elizabeth Marshall
The Hidden Life of Dogs
Houghton Mifflin Company
New York, 1993

MAGAZINES

Dog Fancy
Fancy Publications
Irvine, CA

Dogs in Canada
Apex Publishing
Etobicoke, Ontario

Dog World Magazine
Primedia
Peoria, IL

WEBSITES

American Kennel Club
www.akc.org

American Rare Breed Association
www.arba.org

ASPCA
www.aspca.org

Canadian Centre for Wolf Research
www.wolfca.com

Canadian Kennel Club
www.ckc.org

Delta Society
www.petsforum.com/delta

Digital Dog
www.digitaldog.com

Discovery Channel Online
www.discovery.com

Dog Fancy
www.animalnetwork.com/dogs

Dog Owner's Guide
www.canismajor.com/dog

Dog World Online
www.dogworldmag.com

Guide Dogs for the Blind
www.guidedogs.com

International Wolf Center
www.wolf.org

Minnesota Wolf Alliance
www.nnic.com/mnwolves/

Show Dog Magazine
www.showdog-magazine.com

Wild Canid Survival and Research Center
www.wolfsanctuary.org

Wolf Recovery Foundation
www.forwolves.org

World Animalnet
www.worldanimal.net

·INDEX·

Text references are in plain type; photographs in **bold**; illustrations and charts in *italic*; dog breed profiles in **bold** with asterisk (*).

◆ D · E ◆

◆ F ◆

◆ G · H ◆

ACKNOWLEDGMENTS

The editors wish to thank the following:

Tessa Albanis, Lachine, Que.;
Maryann Burbidge, Royal Ontario Museum and University of Toronto, Toronto, Ont.;
Linda Cobb, Queen of Clean®, a clean little division of Queen and King Enterprises, Inc., Peoria, AZ;
Ross Dawson, Vet Help Inc., Kitchener, Ont.;
Judy Kurpiel, President, International Professional Groomers, Elk Grove, IL;

Jerome Pruet, Nilodor, Inc., Bolivar, OH;
Joanne Ritter, Guide Dogs for the Blind, San Rafael, CA

The following people also assisted in the preparation of this book:

George Constable, Lorraine Doré, Dominique Gagné, Michel Giguère, Angelika Gollnow, Patrick Jougla, Emma Roberts, Rebecca Smollett

PICTURE CREDITS

Larry Allan/Bruce Coleman Inc. 36 (A&C), 39
David Baron/Animals Animals 70
Norvia Behling 33 (left), 36 (B), 46, 55 (both), 57, 63, 65, 71 (left), 77, 78, 81, 89, 94, 98, 100, 103, 110, 115, 134 (right), 140, 148, 161 (left), 170, 175
Tom Brakefield/Bruce Coleman Inc. 19
Jim Brandenburg/Minden Pictures 54
J.C. Carton/Bruce Coleman Inc. 24
Robert Chartier 134 (left), 136
Dani/Jeske/Animals Animals 35
John Daniels/DAN12/Bruce Coleman Inc. 181, 183, 191 (upper), 200 (lower)
Kent & Donna Dannen 32, 36 (D), 38, 95, 127, 128-129, 150, 154, 161 (right), 168-169, 173 (upper), 174, 176, 177 (lower), 178 (both), 179 (both), 186, 190 (all), 191 (middle), 193 (middle), 196 (middle & lower), 197 (all), 198 (upper & lower), 199 (upper & lower), 200 (upper), 201 (lower), 202 (middle & lower)

Richard Day/Daybreak Imagery 5 (left), 10-11, 28 (right), 41, 120
Danilo S. Donadoni/Bruce Coleman Inc. 27
Jim Dutcher/Dutcher Film Productions 2, 44-45, 51, 56, 60
Don Enger/Animals Animals 151
Cheryl A. Ertelt 104, 106, 107, 112, 118, 124, 138, 146, 166 (right)
David Falconer/Folio, Inc. 97
Isabelle Francais 5 (right), 8-9, 37, 72 (both), 101, 142, 152, 180, 185, 193 (upper & lower), 194 (middle & lower), 198 (middle)
Howie Garber/Animals Animals 15
Michael Habicht 43, 117, 173 (lower)
Hamman/Heldring/Animals Animals 34, 48
Jean-Michel Labat/AUSCAPE 162
Gerard Lacz/Animals Animals 6-7, 29 (upper), 50, 61 (lower)
Yves Lanceau/AUSCAPE 184
Tal McBride/Folio, Inc. 64
Victoria McCormick/Earth Scenes 21
Joe McDonald/Animals Animals 42, 53

Karen McGougan/Bruce Coleman Inc. 30
Robert McKemie/Daybreak Imagery 92
Hamblin M. Osf/Animals Animals 24 (middle)
Charles Palek/Animals Animals 47
Robert Pearcy/Animals Animals 29 (lower), 74
Robert & Eunice Pearcy 79
Arthur Phaneuf Jr./Animals Animals 177 (upper)
Fritz Prenzel/Animals Animals 90-91, 171
Maryo Proulx 71 (right)
Barbara Reed/Animals Animals 59 (upper)
Hans Reinhard/Bruce Coleman Inc. 33 (right), 203 (lower)
H. Reinhart/Bruce Coleman Inc. 93
Ralph A. Reinhold/Animals Animals 29 (middle), 59 (lower)
Joel Sartorengs/Image Collection National Geographic 58
Ulrike Schanz/Animals Animals 62
Bradley Simmons/Bruce Coleman Inc. 66-67
Renee Stockdale 23, 24 (right), 26 (lower), 31, 40, 52, 61 (upper),

68, 75, 76, 80, 82, 83, 84, 85, 86, 87, 99, 102, 105, 108, 109, 111, 113, 114, 119, 122, 123, 125, 126, 131, 132, 133, 135 (both), 137 (both), 139 (right), 141, 144, 145, 147 (both), 153, 155, 157, 158, 159, 165 (both), 166 (left), 167, 192 (lower)
Renee Stockdale/ Animals Animals 160
Lynn Stone/ Animals Animals 96
Lynn M. Stone/Bruce Coleman Inc. 12-13
Faith A. Uridel 139 (left), 172, 182, 187, 188, 189, 191 (lower), 192 (upper & middle), 194 (upper), 195 (all), 196 (upper), 199 (middle), 200 (middle), 201 (upper & middle), 202 (upper), 203 (upper)
J & P Wegner/Animals Animals 73
Peter Weimann/Animals Animals 20, 22, 25
Robert Winslow/Animals Animals 16-17, 26 (upper)
Art Wolfe 18, 28 (left), 49
Barbara Wright/Animals Animals 121